THE E-Commerce Book

The E-Commerce Book

Building the E-Empire

Second Edition

Steffano Korper

Co-founder of the E-Commerce Program
Dallas, Texas

Juanita Ellis

Co-founder of the E-Commerce Program
Los Angeles, California

ACADEMIC PRESS
A HARCOURT SCIENCE AND TECHNOLOGY COMPANY

San Diego   San Francisco   New York   Boston   London   Sydney   Tokyo

ACADEMIC PRESS
A Harcourt Science and Technology Company
525 B Street, Suite 1900, San Diego, CA 92101-4495, USA
http://www.academicpress.com

Academic Press
Harcourt Place, 32 Jamestown Road, London NW1 7BY, UK

Library of Congress Catalog Card Number: 00-108252
International Standard Book Number: 0-12-421161-5

Printed in the United States of America
00 01 02 03 04 CO 9 8 7 6 5 4 3 2 1

Contents

11. Get Ready for Wireless!   211

Preface

The past two years has seen dramatic movement in e-commerce. We saw the stock market going up and crashing down. We saw thousands of companies being funded by venture capitalists and going public only to discover "it just did not work." We saw companies spend vast amounts of money to get e-commerce sites up and running only to not realize a profit or even expect profits in the near future. With all of these scenarios, e-commerce may seem like a short-lived trend. Think again! The numbers for goods and services selling online are increasing every day. E-commerce is here to stay and will continue to change the way we do business.

We are now in the second wave, where it is even more critical to understand e-commerce and the business aspects and technology required to implement a successful e-commerce solution. The first edition focused on the main requirements needed to build a Web presence: marketing and sales, vendor selection, and security. What makes the second edition so valuable is that we keep that focus, but now have data to back it up. When we first started working with companies to get online, it was all new. The main goal was to get up and running before the competition did. Now, through experience, we know what worked and did not work in this industry. The idea "if you build it, they will come," for example, has not worked. We saw companies that were successful in the first wave go downhill because they neglected standard business practices such as great customer service, strong fulfillment channels, and target marketing to the right customers. In this edition we have expanded on these areas. Currently only a handful of successful companies have flourished in doing business on the Web, including large corporations such as Dell Computer and Cisco, and Web-based companies such as e-Bay and Amazon. What made these companies leaders, and what drove their success? In this book, we tell you.

The book will be filled with stories and case studies of successful e-commerce start-ups and individuals that have conquered this industry. Having worked with top executives from companies such as Xerox, SW Bell, Alcatel, and Fujitsu, we have direct insights to how these large corporations have moved forward. In addition, we have provided guidance to vast audiences as guests on both local and nationwide radio shows, including CNN, DayBreak USA, Computer Daze, and Business Day, as well as interviews in magazines such *Entrepreneur Magazine.*

In this book, we will guide you through the key elements of being successful online. Think of this book as *The E-Commerce Cookbook.* First, you find out what the right ingredients are. Then you combine the right amounts of each ingredient. Once the entrée is complete, you improve the appearance with garnish. While e-commerce might not be as straightforward as a four-star dinner, there are key ingredients that will start you well on your way. Other companies have been successful in e-commerce, why can't you?

1

Setting the Vision

We have seen e-commerce go through the first wave, with thousands of new start-ups and companies going online. The first wave was the business to consumer market. The second wave was the business to business market. Venture capitalists invested heavily in everything with an E. People started investing based on speculation. Companies went public without a solid management team and products that were based partially on hype. We had not seen such a movement. It was like the gold rush.

You may think e-commerce was just a phase with all of the changes we have seen in the last two years. There were ups and downs in the market, companies filed bankruptcy and venture capitalists pulled back from investments. If you think about it, over 90 percent of new non e-businesses fail in the first year. We knew this would happen; it was just of question of when this shakeout would occur.

E-commerce will remain as a medium to sell products, services, and content over the Internet. Both small and large corporations will extend their enterprises using this Internet medium. The key to the jackpot: *Do it right, right now.* Companies that have done it right have a vision of where to begin, but no vision that would cap their end potential. Like the market, their sight is infinite. They already are ahead.

Market Predictions

Two years ago, Forrester Research, the leading independent Internet firm, released a report containing its first financial estimate for e-commerce. "Internet commerce is transforming the way enterprises do business. . . . By 2002, the value of goods and services traded between companies over the Internet will represent 1 percent of the global economy (approximately $317 billion)."

In 1999, the same research firm increased its earlier estimate for Internet trade fourfold to $1.3 trillion.

Enter the year 2000, and generally accepted estimates top $2.2 trillion by 2003.

This year, International Data Corp. predicts the U.S. online population will reach 137 million—still less than 50 percent of the nation's population.

U.S. online retail revenues will hit $38.8 billion, up from $20.3 billion in 1999, Forrester predicts. About 11 million U.S. households will make their first online purchase this year, bringing this country's total online shopping population to 28.4 million households.

But the online growth isn't just in our own backyard. Forrester predicts that Latin America will see the world's largest percentage growth in annual Internet advertising spending—increasing 137 percent to $121 million. Worldwide Internet advertising, which hit $3.3 billion in 1999, will almost double to $6.5 billion in 2000.

Considering such statistics, it's clear why respected research and development firms now support predictions of a radical new marketplace without physical or fiscal limits.

MARKET SPOTLIGHT

Travel Planning

Even in existing e-commerce markets, the biggest players and ideas may yet be under the surface. Four years ago, showing up at the airport without the old book of check-sized coupons meant serious hassles. Then the e-ticket was born. Show up with your driver's license or passport, and the airline remembers your ticket for you.

The advent of the e-ticket made online ticketing a snap. Suddenly, every airline had seek-and-find Web sites where consumers could browse the weekly specials, peek through travel news and destinations, and then shop for the best fares or itineraries, often comparing several airlines and departure and destination points before arriving at their best possible choice.

Internet travel provides a prime example of the exponential rate at which business can move from bricks-and-mortar to online, adding value to the transaction with the kind of speed, service, and 24-hour convenience available only online.

News released by Forrester Research in April proved how, even after a good percentage of a market's Internet potential has been tapped, room for considerable growth remains. On April 17, Forrester announced that companies eager to save travel and processing costs would start requiring their employees to make travel purchases via Web-based corporate booking engines.

"In 2000, managed travel will already account for 57 percent of the $4.9 billion businesses spend on travel online," said Henry H. Harteveldt, senior analyst at Forrester. "By 2004, however, corporate policy will dictate the spending of 77 percent of the $20.3 billion of business travel booked online."

Forrester further predicted that companies with the largest travel budgets would be the first to shift their travel buying online through the installation of a booking engine, with 70 percent of Fortune 1,000 companies making the migration by 2002. Smaller firms will move their business travel online as well, as secondary companies began to buy seats on high-volume routes in bulk then to resell those seats at a markup to small businesses.

In the end, the online ticketing concept will have exploded into a tremendous marketplace servicing both consumers and businesses, where many airlines and agencies compete for market share.

Defining Vision

It's 11 P.M., and in thousands of dimly lit bars spread throughout hundreds of cities, great ideas for the world's best companies are making their debuts on cocktail napkins. The idea makers get excited—worked up, even—as the profound natures of their napkins grow in the neon glow. They share their napkins with Joe the bartender and call over their billiard-playing friends before folding the napkin carefully and tucking it in their pockets.

The fated napkin may even make it to the office the next day, and the idea could spark similar genius around the water cooler. It gets minds churning. It rallies people to listen, to think, and to evaluate their own ideas. Soon, everyone is offering input on how the idea on that napkin could found the next eBay or Amazon.com. If the creator does nothing, then the idea is worth nothing more than the thin, easily dissolved tissue it's printed on. But, if the creator of that napkin takes action, then that napkin could become a vision.

A vision should establish an excitement and tone for an entire e-venture. The vision must be flexible—good e-commerce companies are in a state of constant flux, always wary of competition and worried about growing stale in a nebulous marketplace.

A vision should establish an excitement and tone for an entire e-venture.

So what happens if you don't have a collection of ideas on crumpled napkins? Do you have the will for an idea, but nothing seems to materialize? If so, then allow us to suggest the following simple prescription for ingenuity:

1. *Take a piece of paper.* Fold it in half, four ways, six ways—it doesn't matter. Put this paper in your wallet, and always carry a pen.

2. *Write down every concept or idea that someone tells you is not possible.* If it simply cannot be done according to one person, then it's another person's e-commerce solution.

3. *Constantly suggest small changes.* If you suggest a change, and the feedback to that change is reluctant—"But we've always done it this way"—then write your suggestion down. You have an e-commerce solution.

4. *Write down everything that frustrates you—everything that takes a little bit more of your time than it should.* Americans are motivated by convenience. They have more money than time and will gladly spend the former to obtain more of the latter. For instance, had you thought, "I hate stopping for directions," you might have

been the first to suggest putting Global Positioning System navigational tools in your everyday Mercedes.

Traditional Business Vision vs. E-Vision

In the end, the opportunity for e-commerce lies in the seemingly impossible. In that regard, developing an ingenious e-commerce idea is not unlike developing an ingenious idea for a neighborhood bricks-and-mortar business.

---

*In the end, the opportunity
for e-commerce lies in the
seemingly impossible.*

In fact, you're making a mistake if you define e-commerce as a different type of business. Instead, it's the same business that's simply based on different tools. Instead of paper and personal time, you use electronics and connectivity. In fact, when people ask us, "What is e-commerce?" we like to say: "E-commerce is lying on the beach, having a cocktail and making boatloads of money."

(This is figurative, of course. Like any business, successful e-commerce ventures are built, figuratively, from the ground up. They aren't get-rich-quick schemes, they're get-rich-slowly schemes.)

Back to our happy e-vacationer:

The beach you're lying on represents how e-commerce operates independent of geography and can serve customers anywhere in the world from anywhere in the world.

The cocktail in your hand represents a dynamic switch in manpower. The goal is total automation—you don't do the work, computer systems do it. Or, at least, they do as much as possible.

The boatload of money, well, that's the reason you're reading, isn't it? If your process is optimized, then you will make money. You will add value to transactions by your fully automated information, product, delivery, and service systems.

Selling on the Internet is not just creating a Web site and thinking, "If I build it, they will come." Companies jumping on the next wave of marketing and sales via e-commerce think outside the box. If they have a current business model, they throw it out and invent a new one. If they lack a business model, they focus entirely on a unique selling point—a single strategy unique only to them that gets the attention of a very specific client pool and can capitalize heavily on the needs of that pool.

Those companies that have been most successful and realized the highest return on investment have approached e-commerce with a specific strategy for building clients, inventory, and revenue. They reengineer their business goals to reflect new revenue streams, they empower customers and provide exceptional customer service, and they consider the implications of joining the first truly global marketplace and economy.

At the same time, these companies don't see e-commerce as an overwhelmingly magical, mystical entity. They just see it as a different way of doing business or as a different business tool.

Remember, as you mull over your vision, using e-commerce as a tool can be very simple. It can be a small way of using e-mail to reach potential customers and clients, or it can be a huge sophisticated payment system.

Regardless of how prepared you are to use e-commerce as a tool, concentrating on five strategies to selling online can help unfold your napkin—and your vision—into a company that can truly take advantage of new methods of doing business.

Rethink Your Revenue

Companies approach e-commerce from every angle, and a variety of methods of selling are being used in the Internet arena. Some companies generate revenue from direct sales, online advertising, subscriptions, and credit card processing. Other e-businesses receive percentages of every online Internet transaction they host. Many businesses earn commissions for matching buyers to sellers. When goods and services are auctioned online, auction hosts may take a flat fee or, like Trade-Out.com, the largest online surplus exchange marketplace, they might take a percentage of every sale.

Companies that generate revenue from direct sales include Charles Schwab and Dell. Companies that generate revenue from online advertising include Onsale.com and Yahoo! *The Wall Street Journal* and

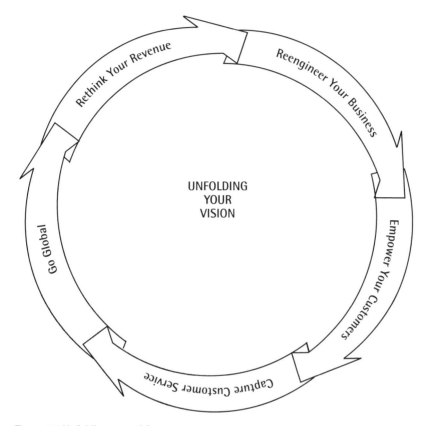

Figure 1.1 Unfolding your vision

TheStreet.com use a subscription-based model. VeriFone and Cyber-Cash, both credit card verification vendors, use transaction fees as a foundation for their businesses. Digital River and Beyond.com use electronic product delivery. Realtor.Com takes a percentage of all real estate agents' commissions whose clients are referred to them via the Web site. Vendors whose sites feature clickthroughs to online stores, including Amazon.com, also get compensation related to the number of clicks they route to other businesses through ad banners and other links.

To start on the journey toward Vision, think about new streams of revenue. As you continue through the book, the vision you have today will be honed by a whole new vision of doing business and increasing your market potential.

Reengineer Your Business

The Internet promises to change your whole manufacturing process, allowing you to communicate instantly with suppliers, partners, and customers on a worldwide scale, yet it requires that old standards of customer service, such as toll-free numbers for instant airing of complaints or questions, be maintained. It is not enough simply to set up Web sites for customers and partners. To take full advantage of the Internet, companies reinvent the way they do business by changing how they produce, market, and distribute goods and how they collaborate within the company and with suppliers.

Reengineering projects may be daunting and complicated. Because e-commerce is not merely a technology-based effort, it requires simultaneously redefining marketing, sales, services, and products. In addition, it requires interactions between suppliers, shipping companies, legal departments, governments, and banks. It requires integration with existing systems either within the organization or between two companies. Successful companies have included marketing, sales, ordering, distribution, procurement, customer support, and trading partners in reengineering efforts. In addition, they have developed solutions that provide complete and timely reference data to company decision makers for marketing, product planning, production, and the logistics planning process.

Empower Your Customers

Unlike a traditional store, the Internet allows customers to take control of any transaction. In the past, the retailer had control. The retailer established what times customers could come to the store, what products customers could choose from after arriving, and what salespeople would be available to answer questions about product features, price, and quality.

Online customers have absolute choice over when and where they want to shop and who they want to contact for information or buying advice. The Web also provides a venue for buyers to have the most choices and find the best prices for any products, be they dust-covered rare books, heavy industrial machines, delicate art prints, or razor-sharp hockey blades. Even in business-to-business commerce, online catalogs, exchanges, and auctions allow companies to purchase goods and services at the best prices.

In other words, where customers traditionally were limited by time, information, and physical distance to find goods at the best prices and quality, today they have absolute control and freedom in tossing out the old models of consumerism. Today, with a click of a mouse, customers can gather information about features and pricing, comparison shop with little effort, seek advice from online news sources or discussion groups, purchase their products, and arrange for direct shipping. They can expect their products delivered to their doorsteps as early as the next morning, and consumers get to slip their shoes on and walk down the hall from their desk chairs. The result is the absolute empowerment of the consumer—the driving force in the sales process.

MARKET SPOTLIGHT

The Automobile Purchase

Consider the traditional automobile purchase, which involves driving from lot to lot; pushy salesmen; right price, wrong color; then right color, wrong price. What if you could pick the perfect automobile from home? This was the vision of Autoweb.com, a company based in Santa Clara, California that went public on March 23, 1999, at an initial offering price of $14.00.

Autoweb.com is now a leading consumer automotive Internet service. With an extensive collection of automotive-related community, content, and commerce offerings, Autoweb.com guides consumers through every stage of the vehicle ownership process, including browsing, buying, enjoying, maintaining, or selling a vehicle. The company's goal is to connect "mass market consumers to real-world buying alternatives through the most extensive network of Member Dealers and other commerce partners, and ensures the consumer experience is faster, better, and easier." As a result, Autoweb.com offers a virtual one-stop shop for automotive enthusiasts, as well as reluctant car buyers. Its site provides links to consumer information, manufacturers' sites, parts suppliers, financing and insurance services, and automotive accessories.

This year, automotive industry consulting firm J. D. Power and Associates reported that consumers saved $490 on average when buying a new vehicle using an Internet online buying service compared to showroom shopping.

Over half the auto-buying public shopped online, and about 5 percent of all shoppers were expected to make an online auto purchase this year.

Capture Customer Service

Customer service plays a critical role in e-commerce, and we'll commit an entire chapter to this later. For now, as you develop your vision, you must remember the following simple fact: Because e-commerce is not a market that geographically captures a customer, merchants must be even more creative in providing value-added services, such as online support, FAQ listings, and follow-up communications and product support.

Your goal will be to automate as many customer service processes as possible by leveraging Web technologies. If this is done correctly, your company will generate not only cost savings but also customer loyalty.

Go Global

Your site should address the global economy. As evident in the statistics mentioned before, the Web is not just for the technology savvy. Women, men, and children of different countries, religions, and nationalities now are lining up to participate in this new market.

But you can't make money from their participation until your company is prepared to go global. Consider the possibility of conducting business online in foreign languages. How easy will your company be to use and interface with? Do you offer a product that will require international shipping and customer service channels? Can you use different interfaces to target members of outlying countries?

The advent of the global economy provides even more opportunities for those of you who *can* think globally to expand exponentially, because the foreign markets will be the last to be fully tapped.

Business Model Basics

Now that you're thinking globally about using new technology tools to capture revenue, capture and empower customers, and provide customer service, you should consider what format or model your business could fit into.

Of course, models should be and are constantly tweaked—and some company has already reinvented each one—but these basic four form the foundation of all types of commerce now online.

MARKET SPOTLIGHT

Business Model Basics

1. Consumer-to-consumer: Think eBay, online newspaper classifieds.
2. Business-to-consumer: Think TheGap.com, CrateandBarrel.com.
3. Business-to-business: Think OfficeDepot.com, TradeOut.com.

Consumer-to-Consumer

This type of model facilitates commerce between consumers, plain and simple. Revenue streams are typically fees for matching buyers with sellers and vice versa. Flat fees and commissions may apply. The most famous C2C company is eBay, the world's largest personal online trading community, which—for a tiny fee—allows consumers to offer their goods directly to other consumers in auction format.

eBay contends that its cocktail napkin moment was a conversation between Pierre Omidyar and his wife, who collects Pez™ dispensers. She wanted to use the Internet as a tool to expand her collection, and Pierre expanded that idea to include a central location for the trade of all collectibles and launched eBay on Labor Day in 1995.

Today, the secret to eBay's profitability is volume—insane volume. Individual consumers use eBay to buy and sell items in more than 4,320 categories, including automobiles, collectibles, antiques, sports memorabilia, computers, toys, books, magazines, music, pottery and glass, photography, electronics, and jewelry and gemstones. Buyers are compelled to trade on eBay due to the large number of items available. Similarly, sellers flock to eBay because that's where the most buyers are. As a result, more than 450,000 items are posted for sale on any given day, and eBay collects a fee on each transaction.

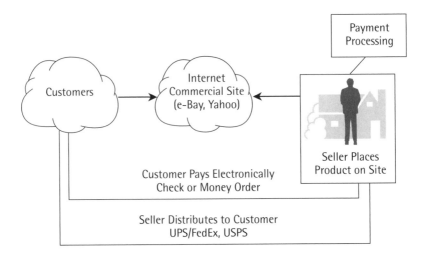

Figure 1.2 Consumer-to-consumer model.

Business-to-Consumer

Think of the business-to-consumer e-commerce as the new QVC of selling your goods to consumers. Traditionally, companies and small businesses employ catalogs, door-to-door sales, print and broadcast advertising, telephone mail order, direct mail, product placement and design, and brick-and-mortar storefronts to sell products. E-commerce is just another tool—a way to sell goods using Web-based technologies. Of course, it's a very powerful tool that transcends traditional boundaries to commerce, allowing customers to hail from any spot on the globe at any second of any day.

Unfortunately, it's also a powerful tool that's enabled almost any business to jump onboard—many that weren't sustainable. This year, Forrester Research released a dramatic report predicting that weak financials, increasing competitive pressures, and investor flight would drive most dot-com retailers out of business by 2001.

To survive in the online retail battleground, Forrester reported, firms would need to redirect extravagant branding investments (which we'll discuss in depth in Chapter 3—Marketing and Branding) into three categories of hard assets: scale, service, and speed. Otherwise, consolidation would steamroll the weak dot-coms, leaving only dot-giants to suck up the majority of an ever-expanding online consumer base.

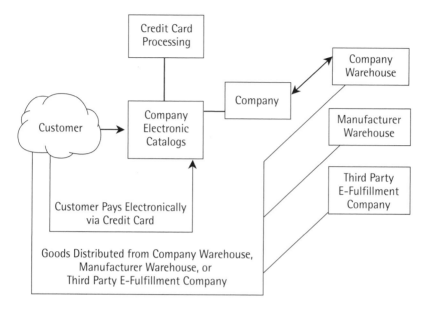

Figure 1.3 Business-to-consumer model.

Forrester even predicted the order of consolidation and demise. First, firms selling products such as books, software, and flowers would consolidate by fall 2000. Second, merchants selling common products at thin margins—drugstores, pet supplies, toys—would fold before Christmas. Finally, they gave e-commerce ventures with significant strongholds in branding longer lifetimes—two years. They'd be gone by 2002. (No one has ever accused Forrester researchers of wearing rose-tinted glasses.)

Whether these dire Forrester predictions hit their mark or not—and we think it's likely—hundreds of e-commerce sites exist, and many currently are highly successful in the business-to-consumer market. These companies search for innovative ways to sell products and services on the Internet. For example, Infoseek allows its visitors to search phone books, yellow pages, and e-mail addresses online. Once you have found the person you are looking for, you can send a card, flowers, or call directly using a specific carrier. All of these transactions happen online.

Markets where convenience, pricing, and selection are major buying influences have done extremely well in business-to-consumer e-commerce. Computer software and hardware retailers have performed extremely well in the e-commerce market. Through electronic commerce, customers have a greater choice of products at lower prices than what they typically would experience while shopping at major retail outlets. Companies such as Compaq and Dell have added millions of dollars in sales per day.

CrateandBarrel.com has taken its brick-and-mortar home goods stores online with great success, as has TheGap.com, which now offers international sales, shipping, and service. Other businesses that have flourished include book and magazine retailers, music and video vendors, and retail florists.

The common market share prediction is for total business-to-consumer e-commerce trade to compose roughly one-quarter of all trade online within three years. The rest of the money will be in business-to-business and digital marketplace models.

Business-to-Business

If, indeed, 75 percent of future Net revenue is in business-to-business channels, then herein lies the greatest opportunity for e-commerce hopefuls. The B2Bs now publicly traded swim daily amid accusations of inflated value and short revenues. Investors jump onboard and offboard faster than passengers on a Carnival Cruise ship docking in St. Thomas for the day.

Regardless, B2B e-commerce ventures will form the backbone of the new economy, which, like the Industrial Revolution, isn't going away. If the publicly traded B2Bs are failing now, it's only because the real gems have yet to be mined and brought to their initial public offering. The biggest opportunities are still in the napkin stage.

Because it offers such tremendous opportunity for specialization and capitalization, business-to-business models can be broken into several distinct forms. First, B2Bs—often called "market makers" by investors—serve either vertical or horizontal markets.

> *Vertical markets* are industry-specific. E-commerce solutions in this category most frequently address the inefficiencies within a given industry, and hundreds of vertical market makers have already emerged in highly scattered, highly populated industries such as steel, chemicals, plastics, and paper.
>
> *Horizontal markets* span across multiple industries by offering common information, goods and services. They provide a venue for

the transfer of goods, and services, including the outsourcing of basic services: data collection and analysis, human resources, and general office operating supplies, to name a few.

Each of the four additional types of B2Bs—online catalogs, auctions, community markets, and exchanges—can fall within either the vertical or horizontal marketplaces.

ONLINE CATALOGS   Perhaps the most basic and well-understood form of B2Bs are the online catalogs, whose main function is to compile inventory from as many suppliers as possible and spur demand for inventory from as many buyers as possible. In the end, the efficiency provided when a hefty volume of suppliers meet a hefty volume of buyers will determine a good deal of their success.

SciQuest, Chemdex, Ariba, and Commerce One are well-known examples.

SciQuest started in November 1995 and sought to use Internet technologies as tools to streamline the purchasing of scientific products, such as chemicals and lab equipment. Today, it is a leading

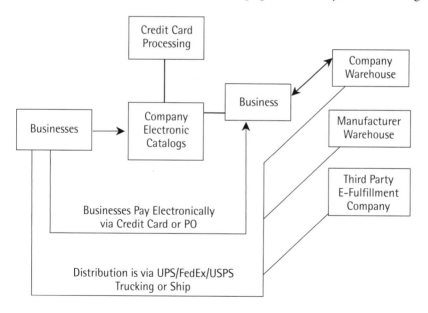

Figure 1.4 Business-to-business model.

business-to-business marketplace for scientific products used by pharmaceutical, chemical, biotechnological, industrial, and educational organizations worldwide. With more than 270 suppliers offering almost a million products, they can easily attract buyers seeking to streamline procurement of scientific industry products.

Chemdex, now owned by Ventro™ Corporation life sciences company, offers e-commerce solutions to researchers and enterprises including an extensive online marketplace with detailed product information on hundreds of thousands of scientific products, electronic procurement to support business workflow and controls, systems integration to interface with third-party and back-office applications (see Chapter 8—Know and Maintain Your Infrastructure).

Ariba developed one of the first true B2B e-commerce applications to fully leverage the Internet. Ariba's first application, Operating Resource Management System (ORMS), automates the procurement of operating resources, which consume about one-third of the typical company's revenue base. In other words, Ariba found and developed the resources to eliminate the largest bastion of company inefficiency. They produce automated work flow at its best.

Commerce One, founded in 1994 as DistriVision and reborn as Commerce One in 1997, developed an application to automates the typically inefficient procurement of indirect goods and services such as information technology, travel, entertainment, professional services, and office equipment and supplies. Its Global Trading Web enables buyers and sellers worldwide to trade in a barrier-free environment, creating new Web portholes and business opportunities for those on board.

AUCTIONS   Auctions provide a venue for the purchase and sale of unique items, such as surplus inventory, used capital equipment, discontinued goods, and perishable or refurbished items. Best examples include FreeMarkets, a reverse auction for manufacturers, and Trade-Out, an auction for surplus inventory.

FreeMarkets creates business-to-business online auctions for buyers of industrial parts, raw materials, commodities and services. Since 1995,

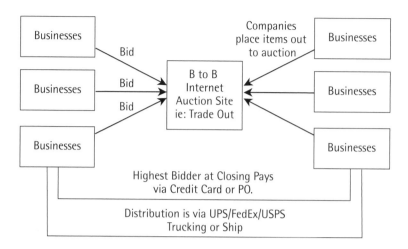

Figure 1.5 Auction model.

they have created auctions for goods and services in more than seventy product categories, including injection-molded plastic parts, commercial machinings, metal fabrications, chemicals, printed circuit boards, corrugated packaging, and coal. In 1999, FreeMarkets claims to have auctioned more than $2.7 billion worth of purchase orders while saving buyers between 2 and 25 percent.

TradeOut leads the industry as an online auction for businesses buying and selling excess inventory and idle assets. The company, formed in October 1998, brings together thousands of companies to buy and sell surplus assets in over one hundred product categories. TradeOut estimates the surplus asset market at $350 billion and seeks to eliminate its rampant inefficiencies through Web-based marketplace centralization.

COMMUNITY MARKET MAKERS   Community market makers bring together professionals with like interests in industry-specific online communities. Information often includes industry-related news, editorials, job postings, and discussion boards. Their goal is to attract a target audience of potential buyers for industry sellers. VerticalNet is perhaps the best example of such community Web sites today.

VerticalNet began as one online business community, Water Online, in 1995. Since then, the company has grown rapidly in size and scope, building ten dynamic, industry-focused Internet communities and environments where business efficiency can be increased as professionals

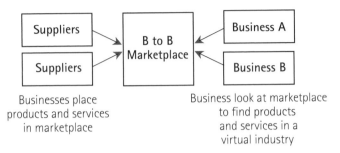

Figure 1.6 Community market maker model.

keep abreast of breaking news and trade ideas as well as goods and services. They have partnered with Internet industry leaders, including Microsoft, IBM, British Telecom, and Internet Capital Group. Advertising is VerticalNet's principal revenue stream.

EXCHANGES, OR DIGITAL MARKETPLACES   Perhaps the most hyped new B2B business models are the exchanges, or digital marketplaces. Exchanges provide a market for commodities, often with high price volatility. Natural gas, telecommunications, or electricity may be bought or sold here. These markets are bid/ask and provide real-time pricing. Exchanges find their revenues in membership and transaction fees. A good example is paperExchange, which used such a business model to bring never-before-heard-of efficiency to the $300 billion global pulp and paper industry.

PaperExchange is completely free for buyers and makes its money by charging sellers a commission for transactions completed on its site, much as a real estate broker might charge commission for transactions completed via his or her office. Trades are private, secure, and anonymous. The site also offers community market-maker elements, such as job postings, industry news, and value-added resources, such as credit and fulfillment systems or logistics.

In the end, your vision will do much to determine which business model best suits your needs, but the decision will not be yours alone. Your customers also will help dictate the model that can best facilitate the fulfillment of their needs. You must choose correctly, or you will not

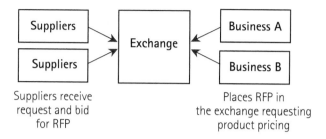

Figure 1.7 Exchange model.

attract the proper audience or be prepared to offer your goods in the best possible environment. You have to know your customers before you can plan the best possible e-commerce site to serve them and successfully vend your solution.

2

Know Your Customers

In 1938, twenty-three friends each contributed $1 to form the Recreational Equipment, Inc., cooperative and to acquire high-quality climbing and camping gear at reasonable prices. In 2000, REI announced that its 1.5 million members contributed to $41 million in global sales online. (Anyone can shop at REI, but its members share in an annual patronage refund.)

Few can dispute the scale has changed in REI's sixty-two years of operation. The company now operates out of fifty-four bricks-and-mortar stores in twenty-two states. Its online store has no borders. In 1999, REI Online had the highest online sales in the outdoor industry, and its customers who shopped online spent 22 percent more in REI stores in 1999 than they had the year before. REI calls its strategy "clicks-and-bricks"—bringing its in-store customers online and vice versa. We call it "knowing your customer."

Before going online, REI strived to ensure that members and customers have access to products and services combining performance, durability, quality, and value. Its sales associates consistently were trained to be goldmines of sporting tips and advice. It backed all of its products with 100 percent guarantees. REI knew this practice worked, and its customers returned because they could depend on an extraordinary quality of products, knowledge, and service.

When REI started its REI Online venture, it took its entire time-tested corporate policy there, too. While online, the customer can access any of the same products, advice, and services. A customer can become a member, read quick primers on equipment, and interact with gear experts. Those new to outdoor sports can access REI.com's educational clinics and learn basic skills, while longtime REI members can check on their patronage refunds or change their mailing address. Because REI.com is accessible worldwide, it also ships worldwide. And its 100 percent guarantee is still firmly in place online.

Knowing its customers liked instant access to any and all outdoor-related news, information, and resources, REI also brought its online product back into its brick and mortar stores via kiosks. While in an REI store, a customer can walk up to a kiosk, access the Internet, instantly learn more about any of REI's products, and order any product that REI carries.

REI didn't attempt to redefine itself or its customer base when it went online; it only sought to expand access to its products. It attracted the same type of buyers, but it attracted them from a worldwide base, not just its stores in twenty-two states. Its sales figures and preeminence in outdoor retailing prove its online success.

Know Where to Start

With sixty-two years of retail experience, REI knew its marketplace and its customers before it struck out into a new wilderness online. For younger retailers and e-commerce ventures, comparable knowledge is equally essential for success.

Because customer acquisition costs are intense online—about $250 on average—e-commerce companies must tailor their entire image and process to attract the consumers who want what they are offering. In fact, the entire content, feel, and look of your Web site should depend entirely on what customers who are interested in your goods and services want online. If you meet or surpass their wants and needs, they will leave your site having had a positive and memorable online experience.

> *Research and target the*
> *right customers.*

Of course, companies like your own must research and target the right customers first. To begin the process, you must understand the demographics and geographies of all Internet users. When you know the base from which you will draw your customers, you then will know what types of customers are available and willing to buy online, and you can identify your best prospects. Then you must learn how these best prospects are accessing your site, how they shop, and what they want online. Only then can you develop marketing and branding strategies and begin to build your site appropriately.

It's a long and tedious process that demands proper commitment. In order to begin successfully, follow these twelve steps:

1. Know who comprises the bulk of Web users.
2. Learn how children will figure into the Web's future.
3. Track how teens spend disposable income online.
4. Watch the heavy hitters—baby boomers and senior users.
5. Know what income brackets dominate the Web.
6. Know what races and nationalities are Web users and in what proportions.
7. Note how gender contributes to online use and sale patterns.
8. Know how much customers in different income brackets are willing to spend online.
9. Know what nationalities and countries are doing business globally online.
10. Research the computer and Web access technology most of your customers will use to access your site.
11. Know what types of shoppers will access your site and how to attract them.
12. Know what these shoppers are looking for online, and give it to them.

Who Is on the Web?

The answer, quite simply, is everyone, regardless of race, gender, age, or nationality, although they are not online in proportionate numbers. The Computer Industry Almanac (www.c-i-a.com) has reported that 327 million people were online this year, and within two years, 490 million people worldwide will have Internet access—about 80 per 1,000 people worldwide. The top fifteen Internet countries will account for about 82 percent of all users.

The Computer Industry Almanac (www.c-i-a.com) has reported that 327 million people were online this year, and within two years, 490 million people worldwide will have Internet access— about 80 per 1,000 people worldwide.

At year-end 1999, the United States led the international brigade with about 110 million Internet users and about five times as many as Japan, the number two–ranked country. Within four years, Internet Data Corp. expects the number of U.S. Internet users to reach about 180 million. A more conservative Net research group, Cyber Dialogue, calls these figures too aggressive and states that about 75 million adult Net users went online in 1999, and that figure will grow to about 109 million in 2003. The real answer may lie somewhere between these two firms' estimates, but both firms agree that the number of U.S. users is, indeed, increasing.

Although the U.S. Internet market is strong, the bulk of Internet access already comes from outside the United States. You will need to think about your business and how it addresses the world markets if you seek large-scale success. Again, the idea is to move ahead of your competitors and to increase your market share. By globalizing your site, you have the opportunity—like REI—to draw greater numbers of unique consumers from a larger consumer base.

For now, adults ages 25 to 44 comprise the bulk of that base and are most likely to access the Internet and buy online. Specialized markets, including children, teens, young adults, baby boomers, and seniors are rising fast.

The Crayon Crawlers

Compared with 2.7 million in 1997, almost 21 million children ages 2 to 12 years will have their eyes peeping upon the Internet in the year 2002. The U.S. government has allocated funds to wire schools, and families are realizing the benefits of computers and the Internet in the learning and research process.

Why do children go online? It's fun. And even though Web sites must now request that children get their parents' permission before giving out information online, the opportunity to build online brand loyalty and familiarity with consumers from their youth should excite online retailers.

An MTV Networks Leisure Time Study showed that children aged 4 to 11 were more likely than the population as a whole to say they enjoyed using computers and the Web. The same study reported that 83 percent of kids use their computers as much or more than they did a year ago. That old family room television only scored 75 percent in that poll.

Clearly, children are and will become more of an online force. The online interests and buying habits they develop while young should carry into their teens and young adulthood.

The 1998 Nickelodeon Online Attitude and Usage Study showed girls aged 7 to 14 flocked to online chat and message boards rather than Internet downloads. Boys the same age did exactly the opposite, while both genders enjoyed playing games and gathering information online.

While the online market for children may have as far to grow as its customers, joining in early may have its advantages.

The Transient Teens

Forget the mall, today the Internet serves as teens' all-purpose hangout—and the Internet mall never closes. To the woe of some parents, teens see the Internet as more than a tool for homework. Online, they talk to friends; explore new communities, fashions, and attitudes; and spend their after-school-job cash.

The half of U.S. teens with Web access surfed the Web about an hour each day in 1999, the Nickelodeon/Yankelovich Youth Monitor reported.

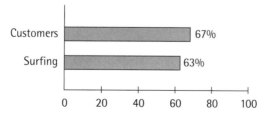

Figure 2.1 Activity on the Web.

Almost one-third spend less time watching television in order to spend time online.

Top Ten Reasons Teens Go Online*

1. E-mail: 67%
2. Just surfing: 63%
3. Play games: 58%
4. Homework: 50%
5. Get to the World Wide Web: 50%
6. Easier than the library: 50%
7. Chat rooms: 48%
8. Learn new things: 45%
9. Instant messaging: 43%
10. Online encyclopedia: 37%

Although they've yet to pack the punch online that they do in the retail market, teens have begun to pick up substantial online buying power. Teenage Research Unlimited, a Chicago-based firm, reports that about 31 million teenagers spent $141 billion in 1998. They spent $58 million of that total online. By 2002, teens will spend $1.2 billion online, and their online population will nearly double, from 8.4 million to 16.6 million.

Most teens buying online buy music, books, and electronic equipment, and they favor the same sites as adults of their same gender. Forrester Research, Inc., reports that the correlation between brand preference and site traffic is lower among teen consumers. For instance, teens who prefer to drink Coca-Cola don't go to the Coca-Cola Web site while they're online.

While they might shun some big-name product sites, most teens probably aren't surfing extremely little-known sites and unadvertised targets. Alexa Internet reports that 70 percent of global Web traffic goes to fewer than 4,500 sites. NEC Research Institute reports that the best search engine can find only 16 percent of about 800 million pages on the World Wide Web. As a result, sites with no branding at all would be difficult to find, let alone remember, even for these transient teens.

*Source: Nickelodeon/Yankelovich Youth Monitor

Baby Boomers Are Booming

Their buying power makes boomers middle-aged goldmines in the retail industry. Now, they're starting to exercise their financial clout online.

Adults aged 45 to 64 surf the Internet more frequently and stay online longer than their younger counterparts, and their heavy spending habits make them one of the most desired markets on the Internet, according to Media Matrix. In 1999, baby boomers made up one-fifth of all online users. The long-considered Internet-savvy 18- to 24-year-old crowd trailed at 17.5 percent of users.

So intense are baby boomers' commitments to Web usage that they access the Web for an additional six days and a total of almost two hours longer on average than the younger crowd. The Web sites that pique their interest include health, fitness, and personal business sites, and they have the income to support investments in their interests.

Income, Race, and the Digital Divide

We've already mentioned that adults aged 25 to 44 make up the bulk of those accessing the Internet and buying online. Among them, tremendous discrepancies exist in Web user demographics. The National Telecommunications and Information Administration reported in a 1999 government study that the most affluent and well-educated Americans dominated the e-commerce scene and that this domination would continue. This administration termed this discrepancy between rich and poor the "digital divide."

Although Web use is increasing among all income brackets, Americans in the top bracket are ten times as likely to be online as those in the bottom, according to the study. Almost 61 percent of households making more than $75,000 per year were online. In contrast, just 30 percent of households making between $25,000 and $35,000 per year were online. Among households making between $10,000 and $15,000 per year, less than 8 percent were online.

Similarly, in 1998, almost half of all college graduates were online, 30 percent of those who had attended some college were, and just 16.3 percent of high school graduates were online. Mediamark Research (www.mediamark.com) reported that 40 percent of all Internet surfers are college graduates, 40 percent have household incomes of more than $75,000, and 63 percent hold white-collar jobs.

As a result, the statistically more educated and affluent households of Asian/Pacific and of white descent have a clear lead in computer penetration and in Internet access rates. However, the discrepancy extends beyond income. When holding income constant, black and Hispanic households remain far less likely to have Internet access. According to the government study, many who aren't online say the cost of computing and Internet service is just too high. Others say they aren't interested in being online. A Cyber Dialogue survey showed that, among the U.S. adult population, those who weren't online saw "no need to go online."

As a result, the government study reported that the digital divide was increasing. It stated that black, Hispanic, and Native American households were more likely to access the Internet from public or shared computers, such as a library, rather than from a computer at home. Black and Hispanic users are 40 percent as likely as whites to have a computer with Internet access at home.

The report did show some hope, however, for a decrease in the digital divide. It stated that, at the highest incomes (more than $75,000 annually), the discrepancy in computer ownership between white and black households decreased by more than 76 percent between 1994 and 1998. Mediamark also noted this decrease by presenting statistics that online usage has grown among lower-income and less-educated groups.

Much as we expect in the bricks-and-mortar retail marketplace, income correlated with the amount spent online, according to an Ernst & Young study. The study found that online purchases rose as income rose. Among people making $100,000 or more, 80 percent spent more than $500 online the year before. Only 29 percent of those making less than $30,000 spent as much online.

Gender Online

Men, long stereotyped the electronics- and technology-loving gender, don't have a huge leg up online. Mediamark Research reports the gap between the number of men and women online has narrowed to 4 percent or less. Once online, men and women both list three favorite sites: Amazon.com, BarnesandNoble.com, and CDNow, according to an Ernst & Young online survey of 1,200 U.S. Internet users. However, after the three top-ranking shared favorites, men's and women's online tastes diverge. Men prefer to seek out electronics, entertainment, and office supplies online. Women click through sites with health and beauty products, apparel, and online greetings.

Even among experiences shared jointly by men and women, such as travel and child rearing, women were much more likely to be found clicking around eToys.com and children's apparel Web sites. Men were more commonly making travel arrangements.

These patterns hold true when men and women stop clicking to open their online wallets and purses. Both men and women buy mostly books, music, and computers online. After the top trio, men buy more electronics, videos, and travel arrangements. Women buy more beauty supplies, toys, and clothing. Men and women also differ in *how* they spend money online. Men were more likely to buy online via auction sites and were more likely to have spent more than $500 online in the previous year.

Geography Online

When comparing one U.S. state with another, Californians led online spending in 1999, racking up almost $6 billion in Internet bills, according to an International Data Corporation study. Texans followed with more than $3.2 billion. The lowest online spenders were relatively rural, less populated states. Wyoming ranked last, with $41.5 million in 1999. The discrepancy was credited to differing population densities, not the percentage of households online, however.

When comparing one international country with another, the density of populations does not correlate with their tendency to go online. The United States continues to lead, and International Data Corp. reports that Americans are 44 percent of the Web population. However, Europe's share of Web commerce is predicted to increase to one-third by 2003.

In fact, the Computer Industry Almanac reports that the popularity of using the Internet is now spreading more rapidly outside the United States, although the United States accounted for 85 percent of all online advertising revenue last year. Although this country is expected to keep about two-thirds of these advertising revenues through the year 2003, by 2005, U.S. Web users will account for less than one-third of all users worldwide. This could result in significant changes is both global advertising and in the language in which content is presented.

Speaking the Global Language

If you conduct business in the United States, conducting business worldwide via the Web has been easy so far. In 2000, more than half of all online business is being conducted in English.

However, as the number of Web users continues to expand overseas, so will the prevalence of conducting e-commerce in different languages. In 1999, about 43 percent of Web users worldwide were non-English speakers—155 million English versus 147 million non-English, according to Global Reach (www.glreach.com). About 96 million Web users spoke in non-English European languages, but the biggest contenders to English were Japanese, German, Spanish, Chinese, French, Korean, and Italian.

By 2005, Global Reach expects Chinese to dominate the Web with 300 million online speakers, compared with English's projected 256 million and Spanish's projected 100 million. Other, smaller players include Dutch, Russian, Portuguese, Scandinavian languages, Polish, Arabic, and Hebrew, with Hebrew being spoken by just 1 percent of 288 users online in the year 2000. Other languages have smaller shares. (See Table 2.1.)

Language considerations in e-commerce are those considerations that address the languages people speak instead of any geographic concerns, such as the specific countries that customers inhabit. It is important todistinguish between the two when considering the additional

Table 2.1   Online Populations (in Millions)

Language	1998	2000
English	107.2	155
Japanese	14.4	21.5
German	13.9	19.9
Spanish	14.2	19.5
Chinese	6.4	15.5
French	8.3	13.2
Korean	n/a	10.4
Italian	n/a	10

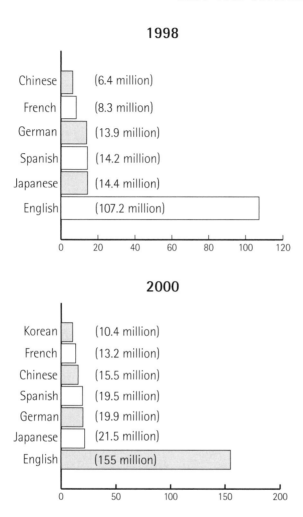

Figure 2.2 Online population growth.

capabilities you might add to your e-commerce site, such as multilingual support or localization. Some countries that your site targets may only require a single language (with English as an alternative). Businesses in most countries outside of the United States provide sites that contain both their native language and English. Other multilingual countries, such as the United States or Switzerland, may require support for several major languages.

Global Reach states that people speaking the same language "form their own online community, no matter what country they happen to live in." For this reason, language and global Internet usage must come into play when learning about and projecting your customer base. Simply saying, "We're a global company," will not be enough. To be a global company, you must both communicate and interact without barriers worldwide. Remember at least to consider conducting business in the native language of your audience, whether you target a single region or a single population.

Web Gurus Are Already Global

The sheer size of the international market could mean tremendous payoffs for ventures that succeed in truly barrier-free worldwide commerce.

TOP 15 NATIONS IN INTERNET USE, 1999*

TOTAL ONLINE POPULATION

1. United States: 110,825,000
2. Japan: 18,156,000
3. United Kingdom: 13,975,000
4. Canada: 13,277,000
5. Germany: 12,285,000
6. Australia: 6,837,000
7. Brazil: 6,790,000
8. China: 6,308,000
9. France: 5,696,000
10. South Korea: 5,688,000
11. Taiwan: 4,790,000
12. Italy: 4,745,000
13. Sweden: 3,950,000
14. Netherlands: 2,933,000
15. Spain: 2,905,000

*Source: Computer Industry Almanac

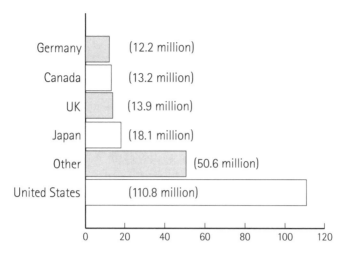

Figure 2.3 Top 15 nations in Internet use, 1999.
*Source: Computer Industry Almanac

By region, a shift in the dominant percentage will continue. In the year 2000, more than 43 percent of all Internet users hailed from North America, according to the *Computer Industry Almanac*. Another 25 percent accessed the Web from Western Europe, and another 20 percent accessed it from Eastern Asia. By 2005, just 30 percent of Internet users will hail from North America. Western Europe will increase to almost 28 percent, and Asia will rise to almost 25 percent. The remainder will be shared by Eastern Europe, Latin America, the Middle East, and Africa. The percentage of users hailing from Africa and the Middle East in 2005 isn't predicted to break 4 percent.

In business-to-business e-commerce, it's not the number of people online, but the value of their online commerce that will determine the viability and profitability of global trade. In 1998, the United States led e-commerce with $37.4 billion in commerce. Japan followed most closely with $2 billion, making other countries competitors, but hardly serious ones. In the year 2002, the United States is predicted to top $409 billion, with Germany and the United Kingdom following with more than $100 billion combined. By then, the top eight e-commerce countries will have totals in the high twenties and teens, as reported by International Data Corp.

Although the market outside the United States is clearly set to explode, if you conduct the core of your business in English or any other one particular language, you must weigh comprehensively the costs and benefits of going global. Localizing a site could be more practical and productive in targeting your consumer than providing additional language options. In business-to-consumer, conducting business in only one language limits your consumer base. In business-to-business systems, your trading partners also will be limited without multiple language capabilities. A cost-benefits analysis will prove essential to your decision, but, in the end, you must make the call.

Companies who chose to include multilingual support and have thrived include Land's End and Dell Computer Corporation.

Lands' End, a catalog-driven global organization, uses its own overseas operations to set up online overseas. The company has separate Web sites for Germany, Japan, and the United Kingdom, each with its own localized address. The German and U.K. sites provide company information, allow consumers to request catalogs, and offer e-mail customer service. The U.K. site also gives information about physical stores in that country. To date, neither site allows online transactions, instead referring customers to the U.S. site. The Japanese site is transactional, allowing consumers to order catalog items and overstocked items. It also offers some features specific to the Japanese market, such as a dictionary of specialized words and phrases used in the clothing industry and links to entertainment articles.

Dell has been doing business internationally and online for years, so it's not surprising that the company has launched specialized sites for dozens of countries. Unlike books or clothes, its products must meet different technical specifications depending on the country. Power supplies and encryption software are just a few of the components that vary by country, as do tariffs. International customers are unable to buy from the main American site, and the order page requires consumers to state that they will not export the systems. But Dell's main home page includes options that take a user to a specialized section of the site with local content, currency, and language tailored to the customer's geography.

It is possible to grow an international customer base without offering multiple languages, especially if your Web site is in English and English is understood by a large percentage of your customer base. In that case, advertising your English-only Web site in your customers' native tongue may be a good idea, even if English is understood by a large percentage of that population.

For the latest information on online demographics, try the following sites:

Name	URL	Description
AllNetResearch	www.allnetresearch.com	A superstore for Internet research
CyberAtlas	www.cyberatlas.com	A reference desk for Web statistics
Forrester Research, Inc.	www.forrester.com	Leading independent Net research firm

Access Issues

Once you know what types of people may be accessing your Web site online, it's important you recognize how they will be accessing it. You already know that Hispanic and black families in mid- to low-income brackets are more likely to access the Web via libraries or other shared sites, while their white counterparts are more likely to have Web access at home. The type of equipment they find to access the Web at any location and the privacy they have while accessing the Web could have a tremendous impact on their ability and desire to conduct transactions and to interact online.

More than nine times out of ten, the connection a customer uses to access a Web site is slower than the connection of the company gurus who originally navigated and approved the Web site design. Nielsen NetRatings announced in March 2000 that 47 percent of Web users have modems with speeds of 33.6 Kbps or slower, and 93 percent connect at no more than 56 Kbps. The 56K modem is about 25 times slower than a high-speed T1 line, which users are more likely to find at work than at home.

As a result, the average Web user's online experience can be classified as moderately to severely inhibited. He or she may be less likely to interact online where multiple graphics increase download times that lag on lethargic, aging modems. A site that requires a plug-in for viewing—even if that plug-in is offered as a free download elsewhere on the Web—may require its customers to spend many minutes downloading prior to viewing the intended site. Unless they're intensely dedicated to viewing that particular site, many customers will just give up rather than devote their evening to downloading. As a frame of reference: Only 12 percent of those online have downloaded the free version of RealPlayer, the most popular plug-in enabling common media streaming.

In the end, a site that boasts the best and most creative designs, replete with multimedia displays and attention-catching visual energy, might in fact be alienating the bulk of its clients if the majority of them hail from the customer base at large. Variances in equipment are expected online, but they can be inconvenient for Web businesses.

Research into your customer base can improve this, but in the end, the most accessible sites are those that can be quickly navigated by customers with the lowest common denominator of computer and Internet access capabilities.

Targeting the Best Shoppers

E-commerce ventures, just like retail stores, will sell their merchandise to three types of customers: the customers who know what they want and go get it, the customers who want to buy something but are undecided, and the potential customers who merely browse the Web. The more directed the shopper, the easier the sale—as long as that shopper can quickly and easily find the product online, get all information on the product, and then complete the purchase. Therefore, Web sites should concentrate on the shopper who came for a product and wants to quickly and easily purchase that product.

The well-known Amazon.com has a small search box that allows its customers to search for any product by title, author, artist, or other applicable data, then quickly lists all available versions, prices, and the average shipping time of that product on the screen. One-click shopping allows past customers to purchase the book or books without reentering credit card information. The directed shopper can access, buy, and leave the site with minimum effort.

However, sites that ignore the less directed, undecided buyers will miss a good deal of clientele. If your site and products attract these types of buyers, you must prepare to develop a storefront that help them easily and quickly browse products by brand, age, manufacturer, category, price, and any other appropriate criteria. These shoppers will buy when they find a product they like, so they should be exposed to as many products as possible.

At the online clothing outlet Bluefly.com, customers can enter their email address and then enter a size or a range of sizes for many different types of clothing, shoes, and accessories. Customers can chose what types of merchandise interest them and what designers and styles they prefer. From then on, they will be presented with their own personalized catalog of merchandise every time they log on. By personalizing the site, Bluefly.com can present its browsing customers screens of images of the merchandise most likely to fit them and to appeal to their tastes, dramatically improving the chances of them finding the products that entice them to buy.

The least directed, browsing shoppers are the most difficult to convert into buyers. Personalized searches, specialty shops, hot-product lists, and online promotions help attract these shoppers, who often will be drawn to a Web site's "glitter," or its look and feel, rather than its content, service, or products. Sophisticated marketing campaigns and online tools such as generalized, all-product queries; chat room and discussion groups; and other promotions can help turn these browsers into buyers, but a Web site should not count on these shoppers to succeed. Targeting a strong and directed base of customers is essential.

Give Them What They Want

As in retail shopping, quality, service, and value are the backbone of e-commerce. A business without those key qualities doesn't make either sense or profits, not in a bricks-and-mortar shop and certainly not online, where a customer can leave a bad experience at one shop and enter another in two clicks.

In addition to quality, service, and value, online consumers demand ease of use, privacy, and responsiveness.

1. Ease of Use   Ease of use is the single most important need to fulfill for your potential customers. Online customers are lazy; many are not technically savvy; and they will go elsewhere rather than tolerate difficulties. Ease of use involves simplicity in access and in navigation

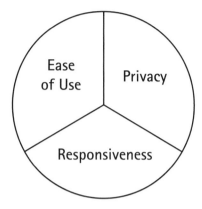

Figure 2.4 What consumers want when shopping online.

online. They don't want to wait more than ten seconds for a site to download. They don't want to scroll horizontally because a Web page doesn't fit in their browser window, although vertical scrolling is OK. They don't want to wade through dozens of features and links that don't interest them. They don't want to try to understand unfamiliar jargon or to read long paragraphs of text. They don't want error messages or too many distractions on the screen.

Instead, online customers want a professional, clear, and concise site. They want to be able to search and get accurate search results. They want the technolingo replaced by easy-to-understand wording. They want to be able to find the right product and conduct the transaction quickly. After the transaction, they want fast shipping and an easy return process, if returns are necessary.

If you are trying to attract customers who are not technically savvy, it is important that the site includes easy-to-access information and simple navigational elements. Plug-ins, elaborate frames, and conflicting color schemes should be avoided. The worst offenders tend to be sites that require multiple downloads of plug-ins prior to offering complete functionality. Keep the interface simple. This includes clear categorization and appropriate color schemes. Focus on the specific demands of your audience.

2. PRIVACY   The anonymity of the Internet is important to most users, and they want to feel secure and confident about entering into a transaction with you before they will become your customers. They fear receiving a hundred junk emails the week after they do business with you. They fear credit card fraud, identity theft, and other abuses of their personal information. Privacy policies should exist and assure your customer that your site systematically protects any personal information given during a transaction.

An important way to support the sale of products or services at your Web site is to include a toll-free phone number. Allow those people that are concerned about Internet security to place their orders (and use their credit cards) with a toll-free phone call. You are more likely to become a victim of fraud by handing a waiter your credit card after dinner than by charging a product online, but Web site visitors remain skeptical. Many sites specify that a third-party certificate authority such as Verisign verifies them. Some sites post the Better Business Bureau logo to show that they belong to the organization. As more people experience online shopping, security fears will subside. For now, do not miss a sale just because your customer is tentative about divulging personal information online.

3. Responsiveness   Customers don't want to feel isolated during or after the transaction. They want to know that real people exist behind the electronic storefront. For instance, if they send an email asking questions or lodging complaints, they want an answer *now,* not in two days. They want a toll-free phone number at which they can reach real human beings with questions or complaints, should an urgent problem arise. In other words, they want more than service; they want responsiveness as instant as the Internet.

Following these steps toward understanding your customer base will help you to target your customers and to develop appropriate branding and targeted marketing campaigns, as well to build a Web site that can most efficiently and successfully give your customers what they want in a positive online experience. A well-targeted consumer base also will keep customer acquisition costs down and help your company more quickly accumulate the mass of customers critical to your Web site's success.

3

Marketing and Branding

After e-commerce companies complete a challenging and tedious process of clarifying their goals and identifying their customers, they must decide how they will define themselves within the new global marketplace. Each must decide what will make them unique and decide how they will use this unique selling point to attract the customers they have already identified.

The process is called marketing and branding. Good marketing dictates how many customers will arrive at and view your Web site, and good branding dictates how they will remember your site and whether they will return.

The principles of online marketing vary little from the principles of marketing bricks-and-mortar businesses. Like the most successful traditional retailers, the most successful e-commerce companies place a major emphasis on marketing their sites. These sites take advantage of traditional advertising channels through traditional media, such as magazines, newspapers, and radio and TV commercials, as well as the new channels available via electronic media, such as banner ads, search engines, and e-mail distributions lists.

In marketing, the goal for any brick-and-mortar or online business is similar: to entice customers to try your product or service, with the hope that their first experience will be a positive one and that they will return as customers again and again while referring new customers. In branding, the goals are brand recognition, customer loyalty, and better word-of-mouth advertising.

Branding online is more than simply moving a print identity to the Web, however. Instead of a good graphic design, your brand in e-commerce is much more influenced by customers' experience on your Web site. Online branding is wholly interactive. Every element of a Web site—its entire look, feel, content, and service—contributes to its branding online. An e-commerce company will have one shot at getting the branding right, or its visitors either will forget about the site or will choose never to return.

Good online branding should result in customers having a positive, unique, and memorable experience on your site. Online branding is wholly interactive. Dissatisfied visitors and customers won't complain; they will leave and go elsewhere.

Depending on your budget, some forms of marketing and branding may be too expensive to deploy immediately. On the other hand, some may prove less expensive than current methods of selling your products and service in the traditional retail marketplace.

Whether you will apply online marketing and branding techniques to your current business or will develop a unique approach for a Web-only venture, think through these steps toward success:

1. Research other companies' Internet marketing campaigns.
2. Know and plan for the acquisition costs.
3. Develop a plan of action.
4. Know what online marketing mediums give the most "bang for your buck."
5. Learn how to best use search engines.
6. Continue to exploit print media advertising channels.
7. Think beyond slogans and logos.
8. Make your online brand usable.
9. Deploy technology that creates a unique definition.
10. Know and use your resources.

Know What You're Up Against

Before you develop your own unique marketing campaign, spend many hours researching those of other companies. Remember what you like and what you don't like. Your potential customers are apt to have the same reaction. Afterwards, dig into these three Internet sites that specialize in Internet marketing: www.adresource.com, www.internet.com, and www.net2b.com.

It's important to acknowledge up front that a few online banners and targeted e-mails will not produce an automatic response. Almost 800 million Web sites now exist worldwide, competing for attention. Companies succeeding online are succeeding after allocating substantial resources to online marketing and branding. They will be your competitors. Therefore, do not commit capital and time to building a Web site if you aren't willing to spend considerable capital and time promoting it after its launch.

A report by Simba Information estimates that online ad spending will reach $7.1 billion by 2002. According to the report, the biggest challenge for Web publishers is to hold the audience captive long enough for potential customers to absorb an advertiser's message. To increase the chance of a message getting through in a medium with more than 800 million Web sites, they must bombard the audience from many angles. Companies use a combination of traditional marketing methods, such as magazine ads and press releases, as well as online advertising at major portals such as Yahoo!, Lycos, and Excite.

For example, sales for Camera World with a site at www.cameraworld.com grew from $1 million in 1997 to over $16 million in 1998. The company sells analog and digital cameras, high-end and standard camcorders, digital camcorders, VCRs, and DVD players to both customers and other businesses. For advertising, they experimented with some banner advertising and search engine listings. However, their most effective advertisements were print ads in *Popular Photography, Petersen's Photographic, Shutter Bug, PC Photo,* and various other camera or electronics-related magazines. They could best target their customers via these popular print publications, which also were industry standards. Camera World also forged partnerships with other online retailers in order to direct the most traffic to their site.

Some companies' most effective advertising is word of mouth. Good word-of-mouth advertising can stem from a relatively small expenditure, as long the resulting marketing is well developed, well placed, and well received. For instance, K&L Wine Merchants at www.klwines.com sells wines and liquor. The company placed ads on Yahoo!, Lycos, and Sidewalk for maximum visibility. But not until September of 1997 did they see sales jump. That month, the company was named as the "Internet's Top Wine Site" by *Money Magazine.* The resulting publicity led to a boom in business, which led to other spectacular word-of-mouth advertising. *Time Magazine* wrote that the site is best for rare and hard-to-find wines, and *Smart Money* magazine recommended the site to its cost-conscious readers as "the cellar to search first."

Larger companies such as Amazon.com Inc., E*Trade Group, and eBay have focused on becoming brand names via the Internet. They have placed emphasis on marketing their site through trade magazines, press releases, banner ads, TV commercials, and other forms of marketing. As a result, they flourish among the other similar e-commerce sites on the Internet. Most of these household names in e-commerce sites were unknown four years ago.

Because they were early movers, their tremendous expenditures in marketing paid off. Other companies undoubtedly made the same expenditures with less effective campaigns, and they struggled to compete with names that were already well known.

Plan for Acquisition Costs

For any new or immature business, the price tag for attracting customers can be staggering. Online, the low percentage of customers who buy, the percent of repeat customers, and overall operating costs can make customer acquisition costs even more breathtaking for start-up companies with small preexisting customer bases. McKinsey & Co. recently completed a study of e-commerce companies and found these companies spend on average $250 to acquire each customer. Unfortunately, only 5 percent of all site visitors will buy something at any one particular site. As a result, companies are spending more on acquisition costs than they are likely to make in profits from an average customer.

Companies spend on average $250
to aquire each customer.

Companies dumping such massive chunks of capital into marketing understand that customer retention will ultimately drive their profit-making capabilities. But customer retention isn't cheap, either. McKinsey & Co. reported that sites spend about $1,931 on site operations for every one repeat customer.

Overall, new customer acquisition costs ranged between $1.92 and $980. The cost of maintaining a repeat customer was between 27 cents and $16,000. The range was attributed to e-commerce businesses' varying maturities and start-up costs, but McKinsey & Co. noted that costs were running, on average, 170 percent of revenues for e-commerce companies more than two years old.

In planning your marketing and branding campaigns, you must prepare for extraordinary expenditures in attracting new customers. If cash doesn't permit, you must prepare to be extremely creative and resourceful and doggedly persistent in your pursuit.

Search Engines and Directories

Georgia Tech University conducted a survey in order to determine how a variety of people discovered specific Web sites. The results showed that the majority of the respondents discovered specific Web sites by search engine listings, by links from other sites or by advertisements in printed media.

NEC Research Institute estimates that about 800 million pages exist on the Web, and a mere 16 percent have been indexed by the best search engines. AltaVista, the largest search engine, has indexed about 150 million pages, according to Forrester Research. HotBot has indexed about 110 million, Infoseek about 75 million, and Excite and Lycos have indexed about 50 million each. Yahoo! is the most popular Web directory, but only about 10 percent of all submitted Web sites ever get listed, and those lucky 10 percent still wait between a month and fifteen weeks for their listing to appear!

Search engines and online directories, both extremely cost effective, are by far the most popular marketing opportunities. They catalog and list your Web site information so that when someone using the Internet searches for information pertaining to products or services that you sell, your potential customer locates your site. Search engines and directories function as listings of your site's theme and content, similar in function to a phone directory advertisement.

As soon as your Web site is up and running, you should list its contents with several of the major search engines and directories or with those that are specific to your industry. Hundreds of lists are available to display the particulars of your company. Some of the major search engines include AltaVista, Lycos, Excite, Infoseek, Northern Light, and HotBot. AltaVista provides 10 percent of all search engine referrals, according to Web Side Story. Yahoo!, a Web directory, provides 44 percent of all referrals.

Although the distinction between the two is often confused or ignored, search engines and directories are the products of two entirely different processes. The processes, described below, also affect your Web site listing when you alter your site. If you, the Web site designer, edit your Web pages, the search engine technology reflects the changes following the next search request. However, editing your Web pages does not automatically update your site listing in directories.

The most efficient means of exploiting these search engines and directories is to list with sites that generate the most traffic. Research indicates that almost half of all Internet traffic generated by search engines and directories is a byproduct of Yahoo! Excite accounts for about 22 percent of the traffic generation, and AltaVista accounts for another 10 percent of the traffic. Because these three venues comprise three-quarters of the Web's traffic generated by search engines and directories, you should begin your marketing work with these three sites. Do not, however, ignore the more difficult to find specialty and industry-specific search engines and directories.

The following seven marketing Web sites can help you better register your own:

1. DirectoryGuide (www.directoryguide.com) features over 350 search engines and directories, each organized into distinct categories.

2. MMG's WebStep Top 100 (www.mmgco.com/top100.html) allows you to list your Web site on the top 100 search engines.

3. The Submit Site (www.thesubmitsite.com) allows you to register your site.

4. WebCom (www.Webcom.com/html/publicize.html) also allows you to register your site.

5. Submit-It (www.submit-it.com) allows you to post your Web site details to more than twenty listings from a central location.

6. SelfPromotion.com (www.selfpromotion.com) also submits sites to search engines and award sites free of charge.

7. If you need further information on search engines, WDVL's "How in the Web Will They Find Me?" (www.stars.com/Search/ Promotion) offers instruction for designing a site, generating publicity, and using search engines.

A majority of the search engines and directories are easy, inexpensive, and effective to use, and it's easy to register your site without professional help.

Search Engines

Search engines create their listings with minimal human intervention. They crawl across the Web to gather information from existing Web pages. Page titles, body copy, and other elements each play a role in the resulting collection of data.

Whenever you use a search engine to search the Internet, you enter a query, then receive a list of sites ranked according to relevance, or how closely content displayed on the site appears to match your query. Search engines follow rules that describe relevance according to the location of words and their frequency. Search engines check to see if the keywords appear near the top of a Web page, such as in the headline, subtitle, or in the first few paragraphs of text. Web pages with keywords appearing in the title are considered more relevant than others to the selected topic. In addition, search engines analyze how often keywords appear in relation to other words in a Web page. Those with a higher frequency are often deemed more relevant than other Web pages.

Every search engine available on the Internet supports both the location and frequency method for listing and ranking sites. Some sites have added other criteria in order to rank pages, just as link popularity and potential "spamming" of the search engine with repeated keywords intended to increase a site's possible ranking among other search results.

In order to have your site ranked effectively, follow six steps:

1. *Make all pages on your Web site accessible within three clicks of your home page.* Most search engines will crawl only three pages deep on your site when searching for content to index your site.

2. *Submit your home page and one other important page, then let search engines begin crawling your site.* Infoseek is the only search

engine that requires you submit every page manually for index-ing. Other sites begin with your home page and then use criteria to determine how far they will crawl within your site. Submit your pages to several of the search engines listed below, because ranking criteria differ between each search engine. In this stage, do not expend too much energy learning the distinctions between search engine ranking methods—the most important goal is to get your site indexed. Just remember that legitimate rea-sons exist as to why searches on different search engines generate different results. After your site is indexed, continue on to step three, and begin tweaking its rankability.

3. *Make sure all your pages have the best possible TITLE and META description and keyword tags.* HotBot, Infoseek, and other search engines give a slightly higher ranking to pages with keywords in their title and metatags. When writing these tags, keep the most important information up front. The first word should be the most important keyword. For instance, if you sell sportswear and running shoes to marathon enthusiasts, the first word of your title should be "marathon," following closely by "clothing," "apparel," "sportswear," and "shoes." "Run" and "athletic" could be other important keywords. The goal is to most closely match the queries of your targeted potential customers, so put yourself in their places. What are they most likely to type into a query? Brainstorm as many possibilities as possible, and put the most likely candidates at the top of the tags. Find more information on revamping your meta tags at www.siteowner.com.

4. *Monitor your submissions, and resubmit often.* Every two weeks is a good goal. An engine that takes your submission one week might drop it a week later. Submitting too often could result in your site being classified as a spammer, but submitting less than once every two or three weeks could keep your site from getting in and staying in the search engine's index.

Some search engines reduce the ranking of pages, or do not list them at all, if they detect search engine spamming. This type of spamming involves repeating a word multiple times in a title or metatag in an attempt to generate a higher ranking for a page.

5. *Get as many other sites linking to your site as possible.* More and more, search engines are using link popularity as part of their

ranking criteria. Excite, HotBot, and Lycos rely on this in rankings and also use link popularity to determine how deeply they will crawl within your site when indexing. In order to gauge link popularity, these search engines determine which Web site pages in their indexes have a multitude of links pointing at them.

6. *Avoid the most common pitfalls.* As we mentioned before, each search engine uses different criteria to determine rank. They also use different criteria when determining which Web sites to exclude. The following Web site schemes might sound like inventive ways to increase your site's ranking, but many search engines will throw your site out if they detect them:

   • Repeatedly using a word on a page or in META tags. Search engines analyze pages to determine whether a word is used too frequently to occur in natural language. A keyword should not be repeated in META tags more than three times.

   • Typing text that is the same color as the page background or very tiny. Although visitors to your Web site cannot see it, search engines will detect it.

   • Using page redirects. Some search engines, including Infoseek, reject pages that take users to another page instantly and without their intervention. If you must use a page redirect, the delay should last about seven seconds.

   • Submitting identical or duplicate pages to a search engine under different file names or submitting them more than once on the same day.

Directories

Directories require that the Web designer or a company representative submit a form containing keywords that describe his or her Web site. Then, when a user enters a keyword for a search, a directory lists those sites whose authors have submitted that keyword as part of the search criteria. The most popular search engine, Yahoo!, uses directories as a method for categorizing information on their site. It's important to understand the process, because Yahoo! generates about 44 percent of all search engine referrals to Web sites.

In order to register your site within the Yahoo! directory, you must search Yahoo! for your Web site and determine whether your site already is registered. If your site is already listed in Yahoo! but you want to change its comments, title, and/or placement, you can fill out and submit a change form.

If you are not already listed, then you must search Yahoo! to determine which category or categories you wish to be listed under. Take your time. Specifying the appropriate category for the listing of your Web site is essential but far from easy, due to the extreme range and number of categories and tiers of information within them.

Click the hyperlink provided by Yahoo! to display a list of important pointers explaining how the directory is organized. Finding an appropriate category for your site is at the heart of this process. Remember that Yahoo! surfers visit each site suggested, and proper categorization on your part helps Yahoo! process suggestions quickly.

If you want to suggest a scheduled chat, live broadcast, or chat area, you should submit it directly to Yahoo! Net Events. If your site is in a language other than English, please check the list of non-English directories to see if there is an appropriate International Yahoo! for your submission. Sites that are in a foreign language represented by international directories are not added to www.yahoo.com.

When you are content you have found an appropriate category, suggest your site from the category you think most appropriate by clicking on the "Suggest a Site" link at the bottom of that category page. An online form will display and request information about your Web site. Provide, among other information, your site's title, URL, and a brief description. Remember that suggesting a site is free, and the ultimate placement or nonplacement of your site within a category will be determined by official Yahoo! Internet surfers. Unfortunately, they are so bombarded by submissions that they often do not have time to look at each one. Yahoo! has begun to offer a $199 express listing service that greatly improves the chances of a business being listed, and the service includes a seven-day decision guarantee.

Otherwise, as we previously mentioned, getting listed by Yahoo! is an extremely lengthy and difficult process, although extremely worthwhile. Yahoo! brings some sites up to half of their online customers. See searchenginewatch.com/sereport/9903-yahoo.html for more tips and advice on how to get listed on Yahoo!.

If you do not wish to spend time researching the search engines and directories listed in Table 3.1 (as well as others) and registering your site, then seek one of many commercial services that will do the work

Table 3.1   Search Engines and Directories

Name	Profile
AltaVista	One of the largest search engines on the Web, a favorite among researchers
Direct Hit	Ranks sites based on the number of hits received
Excite	Offers a medium-sized index and integrates non-Web material, such as company information and sports scores
GoTo	Companies pay to be placed higher in GoTo's listings
HotBot	Favorite among researchers due to its large index and many search features
Infoseek	Offers small to medium-sized index, but includes person listings and company pages
Lycos	A directory model
MSN	Features directory listings and search engine results
Netscape	Offers a branded version of the Excite search engine
Search.com	A branded version of the Infoseek search engine, offering specialty searches
WebCrawler	Offers the smallest index of any major search engine
Yahoo!	Largest human-compiled guide to the Web with over one million sites listed

for you. Both Yahoo! and Express Press can point you in the right direction. Yahoo!'s announcement services provide links to guides and to organizations that you can hire to help publicize your site. User information also is available within Yahoo's Web site. An alternative to Yahoo!, Express Press (www.xpresspress.com), distributes e-mail news releases to hundreds of media editors throughout the world, targeted by subject area.

Advertising Banners

You jump to a new Web site. Not much of interest on this page. As you prepare to jump somewhere else, you notice a small, flashing two-inch by one-inch advertising banner offering a product or service similar to those you were seeking. You click the banner and instantly jump to an entirely new site. The banner has done its job. Banner ads display on a Web page when a potential customer accesses the site. Clicking the banner links visitors to your e-commerce site.

Banner ads are one of the most blatant methods of marketing goods and services over the Internet. Research suggests that online banner advertisements perform as effectively as television advertisements in their ability to increase consumer awareness of brands. In addition, the recall level of a banner ad is equal to that of a television ad. So even if the banner ad cannot coax you to click through it, you still may take its message with you.

The placement of banner advertisements is essential but wholly flexible. Banner ads can be placed on any site your targeted customers are most likely to frequent. For example, if you sell car insurance, you know that when customers buy a car, they immediately must purchase automobile insurance. You should want to place an ad banner on www.automall.com or a similar online car information site, such as www.kbb.com, the official Kelley Blue Book site, which values new and used cars for free. On the other hand, you may decide to place a banner at a bank's Web site, so that when car buyers apply for their car loans they can also apply for insurance. To determine where to employ your banner advertisements, consider site content, placement, and cost.

In order to determine where your banner ads would be most effective, answer the following questions:

1. What are the demographics of my audience?
2. What Internet sites cater to this market?
3. If a visitor searched under the key word "insurance," what sites would he or she reach? Can I place banner ads on those sites to redirect potential customers to my own?
4. What traditional media can I employ to display my banner?
5. Would a banner help my business if I placed it on the Web site of a national or regional magazine?

After brainstorming, choose Web sites with the greatest number of potential clickers and that attract the same types of customers that you are targeting. Or try another good technique: Put yourself in your customers' shoes, and try to surf the Internet as they would. Go to www.yahoo.com, and type in the name of your product, such as "car insurance." Review the resulting list of sites. Visit the ones that seem relevant, and determine whether they offer banner placement. Typically, each site lists a contact number for banner placement opportunities. Some sites allow you to submit banner ads directly from a Web page.

After picking the best sites for placing your banner ads, consider where on their Web sites they will let you place your ads. Although ads can be placed anywhere on a page, they prove more successful when they are appear near the top of the page. The page-leading ad delivers a higher click-through rate than the same ad placed at the bottom of a page.

Once you have considered your options for placing banner ads, take a look at your advertising budget. Some of the major sites charge a substantial fee to place your banner ad on their site, while other may charge only a small fee.

Web advertising is typically sold on a cost per thousand impressions, or CPM, basis. An impression occurs when a visitor to a Web site views a page where an ad is displayed. Banner ads are typically sold in quantities of 100,000 impressions. The cost may vary from $20 to $100 per thousand impressions. Popular sites average $25 to $70 CPM and may offer ad agency discounts and volume or frequency discounts that hover around 15 percent below the gross rate. Because most sites have repeat visitors, 20,000 to 50,000 unique visitors may generate your 100,000 impressions. Most software packages allow you to specify not to display a banner more than once during a single user session.

In addition, some Web sites also sell advertising on cost per click or click through. This type of advertising provides a direct measure of response and ad performance, as you are billed according to the number of times a user clicks on the banner ad. In addition, this information can be easily tracked for analysis purposes. You can determine which sites performed best, and for which categories, search words, and Web pages. Although these advertising arrangements are available, most sites avoid selling this way. Poorly designed ads do not attract clickers, so ad revenue is reduced.

Some Web sites generate substantial revenue from advertising fees. Sites such as Onsale.com rely on banner ads and other ads for a majority of their revenue. You may want to consider using your site as a host for other companies' advertisements. However, in some cases, banner ads for other sites prove problematic. For instance, you may notice that most major department stores, hardware vendors, and other large companies do not place advertising from other sites on their site. These companies want customers to focus on buying their products instead of being distracted by other companies' banner ads.

Depending on your budget, banner ads can be a great way to market your products or services. When assessing their value, consider the sites that act as hosts, the placement of the banner ads on a site, and the expense associated with each.

Online Classified Advertisements

Several sites allow you to place classified ads for free or for a minimal fee. For example, Yahoo! has a classified section to place items you want to sell for free. Other major search engines also provide this service. Classified ads can be placed in several sites. For example, there are thousands of real estate sites that allow you to place ads to buy or sell homes or investment properties for free or at a minimal cost. These sites make their revenue from advertising. Their goal is to get as many customers as possible so they can charge other companies such as real estate, insurance, and mortgage companies higher fees for advertising on their sites. Also, most local newspapers and magazines are online. You can place your ads on these sites at a minimal cost. For example, *The Baltimore Sun* allows customers to place ads in all print editions of *The Sun* and in its searchable online version accessible via www. baltimoresun.com for as little as $19.99 per week. Customers there can search for items and have e-mails automatically sent to them when that item is advertised, which helps you, the seller, further target your customers.

Message Boards

Special-interest boards often serve as listing areas for product or service providers. These message boards tend to address special-interest groups, such as photographers, pet lovers, cooks, or specific industries or businesses, such as biotechnology fields, pulp and paper, or chemicals. Sometimes the special-interest groups are as simple and wide-reaching as parents or college students. One unique board has an ongoing and massive discussion about how aliens are avoiding human contact, although they frequently visit earth. The lesson: Any type of customer you want to target is online and reading a message board.

Boards can be a good supplement to your more visible and costly marketing plans. Yahoo! and Lycos both offer expansive listings of message boards via messages.yahoo.com and boards.lycos.com, respectively. To locate the boards that exist on many commercial sites, especially special-interest sites and the sites of news venues and magazines, search on any of the main search engines for keywords related to your products or services.

E-Mail Newsletters

In order to e-mail newsletters to a well-targeted group of consumers, you must choose registering users as part of your marketing plan and must design your Web site accordingly. If done well, registering users can be a valuable resource for gaining new customers, as well as for ensuring customer loyalty. When users register, you may ask them whether they want to be notified of future developments and special Web events and offers. If your content and products are good, there's a good chance your users will say "Yes."

Once e-mail addresses are acquired, you can send e-mail messages to announce sales on products, to announce new products or events, or simply to promote select products or features. You even can design forms on your site that allow site visitors or customers to select the information they want to receive by category. For example, if you are selling clothing, you could set up your site so that visitors can choose to receive information about specific designers. If the designer announces a new clothing line or if you have a sale on a designer's line, you can send your potential or repeat customer an e-mail message alerting them to the special event.

Ideally, user registration and e-mail technology should be combined for proactive selling. For example, if you are selling gifts, you could set up an online form to allow customers to enter information such as the birthdays of their family members or friends. After processing this information, you could send your customers reminders to purchase gifts before those dates. By capturing the date and providing a service for your customers, you add value to your products and generate more sales.

E-mail messages are a great medium for marketing specifically to repeat clients. Consider an online pool supply store whose best-selling product is chlorine for swimming pools. Customers must replenish their chlorine supplies regularly. The Web site calculates how much chlorine is needed for a swimming pool and how long the chlorine will last and then generates an e-mail message to the customer. That e-mail is an automatic request form for reordering the right type and amount of needed chlorine. The customer can respond by clicking an element, usually a button, embedded in the e-mail message to place the order. Without this functionality, this same customer would have to remember that he or she needed to reorder chlorine and would have to then locate the site on the Web, remember how much chlorine was necessary, and then input the order information and payment method. The improved process ensures convenience for both the consumer and the business—a definite win-win situation.

When e-mail is used as a medium for advertising, targeted e-mail messages are much more effective than scatter shooting. If your customer receives e-mail from your company that is of interest or somehow adds value, you have created a positive experience. On the other hand, if your messages are too frequent or impart little value, you create a negative experience. At best, your message is deleted. At worst, your potential customer feels irritated and denigrates your products or services. When advertising via e-mail messages, consider the guidelines that follow:

1. *Do not reward your customers for registering at your site by spamming them.* Spamming means sending them incessant e-mail messages whether requested or not. To avoid sending e-mail to people who are not interested, ask them on your Web site if they want to receive e-mail from you.

2. *Treat the information a visitor leaves at your Web site as confidential.* Typically, sites inform users that they collect data for internal purposes only, such as analyzing sales trends. Respect the privacy of your visitors. They may become your customers.

3. *If you plan on selling the information that you gather, secure the visitor's permission.* Have you ever received faxes in the middle of the night or useless mail from marketing companies? Companies also employ e-mail messages as a means of selling products and services. Mailing lists are being purchased and sold across the Inter-

net. If you sell visitor information, and it results in annoying or incessant e-mail to your customers, you have violated their trust. Do not trade long-term trust for short-term revenue.

4. *Use e-mail messages to follow up on the sale of a product or service.* Often the only time that you become aware that a customer had a bad experience is after that customer vanishes. Remember that angry or dissatisfied customers do not complain; they disappear. Give all of your customers the opportunity to rate their buying experiences. You can include a questionnaire with a few questions asking about their shopping experience. The information gathered can then be used to improve your site for future customers. Remember that customer service is more critical on the Internet than it is in brick-and-mortar stores. On the Internet, your competitor is just a click away.

Most importantly, remember to include your URL several times in each newsletter you send. Make it easy for potential customers to click to your site and skim its contents. Also, include language on your newsletter inviting recipients to share its content with their friends and colleagues. Don't assume they will share it; plant the idea in their head.

Links on Other Web Sites

Where do your targeted customers go when they surf the Web? Why aren't they at your site? Did they find another related site first? Then use those sites to help them find you!

Search the Web for sites similar to your own, and then negotiate a cross-link with their owners. It may be much more difficult negotiating a cross-link with a site that has a greater traffic level than yours, but you should make an effort to persuade the owners.

If simple negotiations fail, try compensating certain Web site owners for placing your link on their site. Some companies offer a referral fee—a percentage of the sale if a customer purchases a product via the link. For example, Amazon.com offers Web sites 15 percent of a book sale when a customer purchases a book through a link to Amazon.com. The Yahoo! Store includes built-in tools to help you create and manage revenue-sharing links. Use the same strategies as those for banner placement, as discussed earlier in the chapter.

Newsgroups

Newsgroups gained their infamy about five years ago, when the Internet first became popular. At that time, two attorneys advertised their services by posting to every last one of thousands of special interest newsgroups on the Internet. The furor from their postings resulted in angry newsgroup users sending several thousand e-mail messages to these attorneys in complaint. The attorneys had violated the noncommercial nature of these newsgroups. As a result of their scatter shooting, the Internet service provider (ISP) unceremoniously dropped the attorneys as customers.

Newsgroup topics exist to suit every taste and interest, from technology forums to gourmet food to aliens attacking the planet. Posting to these newsgroups is free, but posting randomly can have dire consequences. Use newsgroups responsibly.

First determine which newsgroups might have participants whom you've already targeted as potential customers. Then post information that is informative to these groups. Do not advertise directly on any newsgroups, but concentrate instead on adding value to their discussions. Use information as power, and divulge tips, hints, or other tidbits that create interest among the participants. Pique their interest, and then direct them to your Web site for more detailed information. In order to use newsgroups as an advertising medium, include a signature file listing your uniform resource locator (URL) address.

Discussion Lists

Use discussion lists much like newsgroups to build online relationships with users who share the characteristics of customers you are targeting. Users subscribe and post to discussion lists in order to exchange information about particular interests. Instead of sending e-mails as they are received throughout the day, hosts of these lists compile the e-mails in digest form and e-mail them infrequently.

These discussion lists represent a compelling medium for marketing to potential customers, but, as with newsgroups, we advise that you use discretion. If you choose to participate, then create value for your fellow participants. Ensure that your exchanges are win-win, and your results will be more significant and long lasting.

Table 3.2 lists a few sites that contain bountiful information for both finding appropriate discussion lists and then effectively. Review each of these resources before developing a strategy to target discussion lists and their audience members.

Table 3.2   Discussion List Resources

Name	Profile
Advertising Discussion List	A nonprofit, sponsor-supported discussion group
E-mail Discussion Group Resources	A directory of dicussion lists and resources
Internet Marketing List Archives	No longer active, but archives are available
L-Soft Catalist	The official catalog of discussion lists
List Exchange	A list of discussion lists and resources

Traditional Media

Both small and large companies use traditional forms of marketing to publicize their businesses and to attract people to their site. Onsale.com launched a major television advertising campaign in order to publicize its Web site. Automall.com also uses television advertising to entice visitors to its Web site.

As a starting point for your advertising efforts, place your URL on business cards, customer mailings, company letterhead, e-mail messages, yellow page ads, and any other spot that you currently use to market your company. You may also want to advertise your products and services in the newspaper, in local and trade magazines, in radio and television commercials, in infomercials, or on billboards. As more and more people make purchases via the Internet, online-only advertising will become a more viable method of mass marketing. For now, most successful Web sites advertise via a combination of traditional and online media.

For example, we suggested that a major department store should consider auctioning their products online. We recommended that they place an ad in the newspaper, just as they would for a big sale. This advertisement would attract potential customers to the site because of the curiosity created and for the potential deals available. Once these customers became familiar with the process of the online auction and the resulting auction prices, they were hooked.

As you survey traditional media, consider how businesses support their online efforts and how you can incorporate your online message. In a recent study conducted by the Response Marketing group, Web addresses, used as direct response mechanisms, appeared as frequently as toll-free phone numbers. Among the study's sample group, 58 percent used both Web addresses and toll-free numbers. Street addresses (17 percent), "bingo" cards (4 percent), e-mail (4 percent), and coupons (1 percent) were also used as direct-response mechanisms in the sample. The industries most likely to use a URL in their ads were real estate, water, and delivery services, followed closely by computer businesses, telecom organizations, and office equipment companies. The study also found that URLs are most often used as response mechanisms in the media whose clients were most likely to fit the description of well-educated, well-off Internet users. URLs were used most in computer magazine ads (93 percent), news magazine ads (88 percent), and financial magazine ads (79 percent). The additional cost of adding a URL to your traditional advertising is minimal, but the benefits are invaluable.

Press Releases

Do you offer a product that is unique to your market? Does your product or service add value to the current marketplace? Is it part of a trend, or will it create a trend? The media spotlights products or companies it feels are newsworthy, and the most newsworthy companies are unique or unusual and are spearheading a trend or are part of a larger trend that will affect its audience. Develop a news angle that positions your company within these newsworthy categories.

For instance, are you the first company to offer a particular service online? Put it in the press release. Are you the first in the area? That's good, too. Are you changing the way a certain industry traditionally operates by going online? Now we're talking trends, and trend stories have weight. What will the media's readers, listeners, and viewers gain by learning about your site? If your site could have a wider impact, then it's more important.

To get your name swimming in media circles, develop a compelling reason that your site stands out from its competition, and then send press releases to your trade magazines, local newspapers, online magazines and business magazines. Call ahead to get the name, fax number, or e-mail address of the editor or reporter you will fax. The media are bombarded with thousands of story pitches and press releases each day. Getting it to the right person is half the battle.

In addition to sending press releases to editors and reporters, some Internet sites allow you to submit and post stories directly that might be of interest to their readers. As with any marketing products, make sure the quality of your press releases matches the quality of the image you wish to project. Because you are sending the releases to writers and editors, spelling and grammatical errors are as deadly as factual errors. If you are not up to the task of writing a perfect news release, hire a marketing company to help. It's worth the investment; if a newspaper or magazine editor finds your site compelling, so will his or her readers.

Just Give It Away!

Glittery giveaway promotions work and are a popular way to entice new Web site traffic. TheKnot.com and other sites targeting brides-to-be commonly feature banner advertisements for giveaway promotions that include free honeymoons, wedding-day jewelry—even free bridal gowns for the big day. Women.com has an entire page of links to "free stuff," including coupons and sweepstakes. The page gains credibility by showcasing sweepstakes winners. *Sports Illustrated*'s online site featured a banner ad to offer $20 off a $50 purchase in a Dick's Sporting Goods online sports merchandise store at dsports.com.

Advertise your promotion in traditional media as well. No matter what the medium, request that online users register in order to be entered in the giveaway, or require them to register in order to claim their prize. It's another way to gather precious e-mail addresses from visitors that will help you send targeted e-mails to a good potential customer base later.

Branding for Success

Is Kleenex the only form of tissue? Is Band-Aid the only form of adhesive strip? Is Coke the only form of cola? Branding is a powerful tool. Trademarking is something you do to the public, but branding is something the public does to you.

Trademarking is something you do to the public, but branding is something the public does to you.

Branding is the culmination of customers' experience with your Web site. It's a relationship that must be built and maintained with the goal that people will think of your brand first when they think of purchasing a particular product type. Online, this experience is much more than a logo, name, or a trademark, although these elements of advertising should capture the unique ideas that allow your company to add value.

Online, a brand must be multidimensional and must encompass site design, overall look and feel and an effective logo, as well as a positive customer experience. A recent industry study states that the top one hundred e-commerce sites spent an average $8.6 million each in 1998 to build their online brands and attract customers to their Web sites. Even the smallest companies in this group, those with one hundred or fewer employees, spent an average $2.2 million each on marketing, while the largest companies in the group averaged $21.4 million. The top five marketing budgets in 1998, according to the report, belonged to Amazon.com ($133.0 million), E*Trade Group ($71.3 million), BarnesandNoble.com ($70.4 million), CDnow ($44.6 million), and Ameritrade Holding Corporation ($43.6 million).

These companies are willing to spend tens of millions on branding in order to be the frontrunners and the Internet's brand names. Books online? Amazon.com. Trading online? E*Trade. Compact disks online? CDnow. The principals know that if they brand early in the e-commerce wave, they will capture future markets. Remember the statistics; revenue will reach $2.2 trillion or more in the next four years. Capturing and maintaining a small percentage could prove lucrative long after the initial marketing investments have been paid off.

Establishing a strong brand involves the following three steps:

1. *Know your customers.* This is step number one in most areas of business. Remember, you may develop your product, your Web site, your marketing scheme, and your system of customer service, but only your customers can give you your brand. Remember also that your best bricks-and-mortar customers may not be your best online customers. Therefore, the same branding techniques may not apply to both the traditional and online aspects of your business.

2. *Keep your brand convenient.* Use Internet surfers' laziness to your advantage, and appeal to them through a simply designed site that allows for easy, near-thoughtless site navigation. A customer who does not want to browse—just get what he or she came for and leave—should be able to do so. Poll your customers to determine the usability of your site. If it's not good, improve it in order to improve your branding.

3. *Use technology to add functionality.* Use new technology interfaces to add value to your site. Integrate your Web site with back-end databases, and add customer services, such as real-time stock quotes or stock portfolio tracking. Another technology-enabled service includes personalizing Web sites, shopping catalogs, and news pages so that users do not have to remember passwords or search for the information or products they deem most important or desirable. Make your site a convenient tool that adds value to users' day-to-day tasks, and they will return.

Marketing and Advertising Agencies

Considering the breadth of research and resources required to market and brand online, as well as the magnitude of related expenditures, you might want to turn to the professionals for help. Marketing companies can either support your marketing efforts or develop entirely new marketing concepts tailored to the new global economy. Before you employ a marketing agency, do some preliminary work yourself using the marketing strategies described in this book. This knowledge and background will help you to ask appropriate questions and get your best results.

Most importantly, stay involved in the creative process of developing your company's marketing and branding strategies. Beware of a marketing agency that hands you a template and then tells you *"This is the way you must market."* While some standard methods of marketing exist, each business is unique. The company you select to support your marketing and branding efforts should develop creative ways of conveying that idea to your customers.

One marketing agency, Advanced Marketing Corporation (www.theultimatepromotion.com), serves as an example of the type of support you can expect. Their service, Ultimate Promotion, helps boost your Web site traffic. Their Traffic Building Package submits your Web site to more than 900 search engines and directories quarterly and submits your Web site to over 500 popular links pages. In addition, Ultimate Promotion advertises your Web site with 5,000 displays of your banner ad. Each of these services generates visitors to your site—visitors you can then turn into customers. Additional information describing the company is available at their Web site.

Table 3.3 lists companies that can assist with marketing and advertising efforts. See also www.adresource.com.

Whether with a professional company or with your own e-commerce team, invest in marketing and branding. Combine traditional selling methods with those offered online. The goals are to get as many people as possible to your site at least once. After they arrive, a good impression of your site, its content, and its design will help compel them to make a purchase. After this initial purchase, then you will focus on the customer service channels that will keep them coming back and buying again and again.

Combine traditional selling methods
with those offered online.

In the end, a good overall customer experience can make your branding infallible, even when new competitors arise. When you reach your goal of creating and maintaining a popular site, the result is radically increased sales via word-of-mouth advertising and increased revenue generation by selling other sites' ads.

Table 3.3   Marketing and Advertising Agencies

Company Name	Profile
THINK New Ideas	A full-service marketing and communication company
Beyond Interactive	Features media planning, buying, and campaign management
Ad Up	Offers site and banner design, as well as marketing consulting
Online & Multimedia Marketing Group	Offers e-mail marketing, Web development, and offline promotion
K2 Design	A full-service interactive firm
Eyescream	Provides online media planning, placement, and strategies, as well as creative development for banner and advertising campaigns
USWeb/CKS	Helps clients define strategies and build their businesses
Deadlock Web Design	Web site promotion and design

4

First Impressions
May Be Everlasting

Click on the American Airlines site at www.aa.com. Look at the whistles and bells! Everything you could possibly want for travel, accessible from a host of tabs, pull-down menus, and links. It's all there, isn't it? Or is it? You'll spend a while familiarizing yourself with the site to see, and you'll have to log in to use all of the site's services. The links are effectively organized, and the site is entirely professional and attractive, but it will take a while to get what you want.

Now go to Southwest Airlines' Web page, www.iflyswa.com. You'll find a much simpler home page that loads quickly and is easy to skim and to navigate. The logo is bright and obvious. Only a few internal links are listed, but they are clear and easy to navigate. A year ago, this page was one large graphic that looked and felt like a counter in an airport. All of the possible links and applications within the site weren't initially obvious, but physically, Internet users who had flown on an airplane before knew how to use the site. With both sites, Southwest Airlines had an effective Web presence that could be summed up with the following: Simple. Tactile. Easy to navigate. Efficient. Memorable. *Good e-commerce.*

The overall design and mission of a Web site creates the first impression for visitors. A well-designed Web page will balance unity and variety and will support the business's message without overpowering it. Compare building a Web site to designing it: Building the site is the process of building the customer interface for an entire e-commerce

Figure 4.1 How to create an efficient Web presence.

system with architecture that provides automation for process efficiency. The goals are cost-effective sales and ease of use that increase customer satisfaction, retention, and loyalty. The more complete the automation, the more efficient the site in achieving these goals.

Designing the Web site takes these principles and adds marketing and branding. Because the success of e-commerce so heavily depends on efficiency, it's no secret that the key to good Web site design isn't glitter, bells, and whistles; it's well-defined simplicity that provides an effective storefront.

Follow twelve steps to a simple and efficient Web presence:

1. Focus on the essential element: speed.
2. Use wide, consistent site design to jump-start navigation.
3. Enhance navigation with good link design.
4. Make sure your site is reliable.
5. Keep your overall look clean, professional, and consistent.
6. Make graphics and layout reflect your business goals and image.
7. Keep text readable, and make it enhance your image.
8. Make fonts enhance a site—not detract from it.
9. Make it easy for visitors to skim pages.
10. Make your visitors feel confident.
11. Use your resources: professional design agencies.
12. Revisit your vision statement and goals.

Focus on Essentials

When users click onto a Web page, speed is the most significant factor determining whether they will stay or leave. In fact, online at www.cyberatlas.com, find the CyberAtlas/Watchfire Quality Test. Cyber-Atlas, the leader in compiling Internet research on the Web and Watchfire, the leader in e-business quality analysis, have teamed up to track the performance of the Web's twenty-five most popular sites, and they give each a ranking based on criteria including the number of inoperable links and other Web page blunders. But the test's two top criteria are:

1. No pages over 50 kilobytes in size, including graphics.
2. No pages that take more than 20 seconds to download with a 28.8 K modem.

Most Web surfers agree that speed is extremely important on the Internet, and most consider the biggest problem of the Internet to be Web sites that take too long to download. Graphics are the number one traffic-killer when it comes to Web site speed. The higher the byte size of your images, the slower your Web site will download. It doesn't matter how spectacular the graphics on your Web pages appear if Web users' patience expires before the images can load. Of course, it's important to use a few graphics to entice surfers, Web site viewers, and potential customers. However, it is possible to keep good images on your site that are no larger than 20 KB. In addition, images 3 to 5 KB in size will download almost as fast as text. Each additional 2 K adds about one second of download time on a 56K modem.

The ability of graphics to load quickly depends on both color and size. The fewer colors, the faster the picture will load. When designing photos or other graphic images in the popular Adobe Photoshop, the default setting exports an image to .gif format, which defaults at 256 colors—a surely slow download! Use the "adaptive" setting and preview the image to export the .gif with as few colors as necessary.

Other design programs will convert your images to .gifs differently. The key is to always be conscious of colors and to use as few as possible to still retain a good image. Also, make your images seem faster by saving them as interlaced images so the viewer will at least see something other than an empty box during download time. Speaking of empty boxes, always include alternative text for your images, too, so that the viewer will know what's coming and need not wait before moving on to different pages in your site. Loading a large image on to a Web site then scaling it down to the correct size for your page also is a mistake. The scaled-down image will maintain the byte size of the larger image and will take much longer to load. Determine the necessary dimensions for your Web page, and save your image as a .gif at those dimensions before you upload it to your site. If you want a large graphic on your Web page, try slicing it up into smaller images and putting it in a table in order for the image to download faster. Find the free program to dice images into more workable piece at http://www.ziplink.net/~shoestring/dicer01.htm.

Easy Is Everything: Use Wide, Consistent Site Design

If your Web site loads quickly, your next challenge is making it inherently easy to navigate. One of the first elements affecting a site's navigation is how the pages are laid out in relationship to one another. Another is consistency of design.

Web sites can either be narrow or wide in design. A narrow site means that all information on the site is accessible from many clicks via a few initial links. As a result, your site is narrow and deep. Users accessing a narrow site either have to click through many pages of information on your site in order to find what they are looking for, have to search the site to find what they are looking for, or have to find and view a site map of the site in order to find what they are looking for.

If a site is narrow, having a search element accessible from the site's main page could be important. In any case, Web surfers are likely to give up and move on to another site rather than dig through a narrow site.

Another solution is a wide site, where all of the information or categories of information are accessible from the site's main page. This design allows Web users to access the information they seek without clicking through many pages. Research shows Web users prefer wide sites, which allow them to work faster and more accurately.

Visitors work faster on wide sites because they have to make fewer decisions. From one page, they scan a list of choices. Even if the list is long, they only have to choose once. On a narrow site, they would choose from a sequence of short lists spread over several pages. They would have to make one choice after another. That takes time.

Visitors work more accurately on a wide rather than a narrow site, because in a deep site they are more likely to get lost among the many layers of information. They may not remember where they saw something that interested them as they wade through the site. They may not be able to find something they deemed important after the fact. Ever forget where you parked your car in a multistory parking garage? Or forget what level of a multilevel mall you entered on? The same principle applies.

Wide site design also can help a site's ranking on search engines. As we mentioned before, many search engines will crawl only three levels deep to find and index information on a site.

Consistency of design will enhance a site's navigation. For instance, navigation bars are most commonly placed on the top and left-hand sides of a site, because Americans are accustomed to reading from left to right and from top to bottom. Americans expect to find navigation bars at the top and left-hand side of pages, so they will look in those locations first. On an extremely long page, a short navigation bar at the bottom of the page is helpful and common. Web users will know where to find and how to use navigation bars in these places.

Consistency of width also is important. Web users are used to scrolling from top to bottom to find all the information on a page. They are not, however, accustomed to scrolling from left to right. Therefore, information on a Web site that is too wide for browser windows will not be as widely noticed and accessed, and a too-wide Web page will damage the ease of navigation for the site.

Once you develop a site design you are comfortable with, use certain key elements such as navigation bars, content layout, and standard graphics or logos as templates for your site. If you place these elements in the same spot on every page, visitors will quickly absorb the layout and will know where to look for the information they seek.

Present Interior and Exterior Links Well

When deciding how to present internal and external links on your site, focus on removing the guesswork for visitors and on enhancing navigation. Think of the weather page in your local newspaper, where the column of forecasts is kept for major cities nationwide. Any major city will be there, and the cities will be listed in alphabetical order. The list is the same every day, although the weather information (highs, lows, and precipitation) will change.

Web site navigation requires similar design. From a list of choices, visitors should know exactly what they would find from each keyword. The meanings of links should not overlap or be difficult to decipher because of poor or difficult language. In addition, all links should be in the same place on a page, prioritized from the visitor's perspective. If a site has many internal and external links, these links should be grouped into categories to help speed the visitor in finding the appropriate link. Links should not be embedded in text, scattered around the page or associated with graphics. Two links should not go to the same place, either. Redundancy adds confusion.

Reliability Counts

Another category that influences rankings on the CyberAtlas/Watchfire quality test is reliability. Sites with links that have broken internal and external links get a ranking much worse than their competitors. Sites that are not always up and running will damage their consumers' confidence, as well as hinder Web users' abilities to access and conduct transactions via their site.

Because reliability is facilitated greatly by your systems architecture, we will revisit this topic in our architecture chapter. For now, plan on designing your site so that, if a link is missing or the site does temporarily go down, visitors can easily find the means to access your company or a company help desk through another route, such as an e-mail address accompanied by a quick-response promise or a twenty-four-hour, toll-free customer service line.

Tackle Your Overall Look

Your Web page, especially its home page, is your online storefront. This one page will create an immediate impression on each of your visitors, and it will determine whether the majority of visitors stay or go. Overall, this first page should look clean, professional, and attractive, much like the lobby of an office building.

A clean site has no broken links, no missing page title, and no missing graphics. It is uncluttered and carefully uses banners and links. It's not swamped with advertisements. Banners work well if used sparingly and in appropriate places, but they can detract from a site if their placement is inconsistent or these banners are too heavily used.

A professional site closely mirrors the image of the company it was designed to portray.

A professional site closely mirrors the image of the company it was designed to portray. No one definition of "professional" exists. A professional image for a Harley Davidson dealership will vary greatly from a professional image for an accounting firm. This is OK. However, the general rules of business apply. A site should not be flashy, gaudy, profane, or otherwise offensive. A professional site will be tailored to attract the trust of its clientele.

An attractive site combines these elements, along with the elements of color. Colors should be subdued and attractive. Bright, neon-green headlines or navigation bars scream "Amateur!" Use colors that enhance your design but do not cause it to jump off a page.

Whether you're choosing colors for a graphic image, a background, or for colored text, give some thought to what types of colors are appropriate for your site and the image you're trying to convey. Whole courses are taught on the proper uses of color and how to evoke various emotions with appropriate colors.

Bright primary or secondary colors such as blue, red, yellow, orange, green, or purple are loud, happy colors. They are seldom, if ever, used by stodgy corporate sites, but kids will love and react to them. If your design is mostly black and white, then such a bright color could be used to call attention to certain elements, such as a short bit of text. Remember that these bright colors seldom look dignified.

Darker shades that contain a large proportion of black are much more dignified. A dark blue, hunter green, or brick red can add life to a black-and-white page without overwhelming it. Try using colors like these for headings or for other small elements, such as navigation bars.

Other important shades include pastels and earth tones. Pastel shades are laid back and undemanding. A pastel shade is often a good choice for a background, if white or very light gray isn't appropriate. Earth tones such as light tan or brown are organic, unobtrusive, and popular. They can be used in either dark or light shades, and they tend to contrast well with primary colors. Regardless of what colors you pick, use them sparingly to create visual interest on your site.

While you're thinking about which colors would best serve your interests, remember that working with colors on a computer screen is quite different from working with colors for print. While print applications usually use a cyan, magenta, yellow, and black (CMYK) color model, the Web uses a red, green, and blue (RGB) color model. This model uses three numbers from 0 to 255 to represent each of three colors. However, Web browsers can display only a small subset of the 16 million possible RGB color combinations, and the browsers' subsets are not consistent. Therefore, Web designers stick to a Web-safe color palate to ensure that visitors to their sites will view the site in the colors they designed their site with. If designers use a color that's not in the Web-safe palette, a user's browser may substitute another color. Netscape navigator and many other application programs have a predefined set of colors or a color table that they use to display images in HTML pages. If a color used in an image is not found in the color table, the browser will either substitute a color from the color table that is similar to the color specified in the image, or it will dither the colors to approximate the desired color. Find a good source of information on colors and color combinations at: wdvl.com/Authoring/Graphics/Colour/.

Use even more caution when considering colors for backgrounds. Busy backgrounds can make text impossibly hard to read and can detract significantly from other content on a Web site. Although a well-chosen solid color background can set the mood and give a site a unique identity, choose carefully. Color is powerful, and different colors will have different connotations for your visitors. For instance, a white background implies a no-nonsense business attitude, while a black one evokes an air of creativity and mystery. Red and orange excite the senses and increase heartbeat. Blues and greens are more restful. Yellow reminds us of energy, sunshine, and flair.

Compare www.washingtonpost.com with www.beyondideas.com. *The Washington Post*'s online site uses white backgrounds and careful, subdued colors that reflect its staid corporate image. Beyondideas.com is a Web site creativity company. For those with flash technology, its site begins with bright colors, moving sights, and sounds. However, after the introduction, the Web site reverts to its nonflash version, where professional layouts and no-nonsense white backgrounds are the rule. As a result, Beyondideas.com has presented itself as a creative company that's still a *company*. From its Web site, you walk away impressed by the company's image: a group of creative *professionals*. This is the power of good Web site design.

For each example of professional-looking Web design you find on the Web, you will find many more examples of amateurish design.

Avoid the following pitfalls:

1. Busy, bright backgrounds.
2. Busy, complicated graphics.
3. Bright, elementary-looking headlines and navigation bars.
4. Carelessly or overly used frames that clutter a page.
5. Extra long pages.
6. Under construction signs.
7. Visual clutter: too many banners, buttons, and pointless animation.

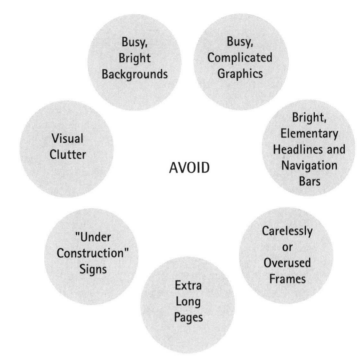

Figure 4.2 No-no's in Web design.

Use Graphics and Layouts Well

We've already discussed that too-big graphics and too-intricate layouts are hallmarks of poorly designed, slow-to-load pages. However, well-designed and well-placed graphics can enhance good layout to relay an important corporate image in an instant. The graphics and layout of your home page contribute heavily to your first impression.

Make sure your layout and design are symmetrical. Think about balance of graphics, images, and margins on a page. The page should be visually pleasing, not off-kilter. Have someone else edit your site for symmetry.

Also concentrate on the quality and appropriateness of your images. If you have a serious site, you don't want cartoonlike images on your site. If your company image is more playful, you will want images that reflect this. When using graphics in layout, remember that any graphic that has a directional aspect should be placed on the page with the intent of drawing visitors' eyes in toward the page. For instance, if you have an airplane on the top right-hand corner of your site, make sure its nose is pointing downward and in toward the rest of the page.

Keep Text Readable

Your text should say the same thing to your visitors as your image. Eliminate discrepancies between what you say in print and what you say visually. If you are sending conflicting messages, neither will be received. Choose words that are as good for your company as your logo. A logo might communicate instantly, and text may take longer for visitors to consume and to digest. But do not assume that text is secondary. Choose your verbal elements carefully.

Your words must stand out on your page. Surround them with plenty of white space. Words crowded by images will make your site visually difficult to digest and also will decrease the probability of visitors actually reading the words. If you have large blocks of text for readers, remember to control your line length.

Long an important tool for print designers, controlling line length greatly increases text's readability. Imagine a newspaper page without columns. The slab of text that would cover the page, margin to margin, would seem overwhelming and imposing. You, as the reader, would be unlikely to begin reading, let alone finish an article. For this reason, newspapers and magazines divide their text into columns. Those columns break up text into digestible pieces.

Keep this in mind when placing your text on the page. Control your margins. See www.omaha.com/OWH, the online site for the *Omaha World-Herald* newspaper, which uses blocks of black and muted colors to help draw readers' eyes onto text kept solidly within easy-to-read margins.

As important or more important than the appearance of your words are the words themselves. Poor spelling, bad grammar, and careless punctuation seriously detract from the appearance and readability of a site. They hinder a reader's ability to focus on content and destroy a reader's ability to respect the site as authoritative or professional.

In addition, such mistakes reflect poorly on the Web site owner and the owner's product. If you are not a gifted writer, take steps to improve your skills. Have someone proofread and edit your work, or hire some-one to write your pages for you. Do not ruin a well-designed site with poor text.

Make Fonts Enhance a Site—Not Detract from It

Select one font for all of your headings and subheadings. Select the same font in a smaller size or a different font for your text. Use boldface, italics, or size to distinguish important elements, but keep your fonts consistent. Too many fonts on a page will make the page visually confus-ing and amateurish, just like using too many colors.

The purpose of fonts is to convey information. Well-used font styles will allow your visitors to glance at your page and to take in all the key points.

Remember that many users can or will specify via their own browser's preferences which font they will view your site in. Do not rely too heavily on a specialty font to convey meaning on your site. Instead, concentrate on using white space, different sizes, and different styles of the same font in order to draw attention to certain elements. Small bits of text can be presented in color, if you wish to draw attention to them. In general, the standby black text is best.

Make Your Pages Easy to Skim and Revisit

Dividing your pages into easy-to-digest bits is essential. Make sure headings, subheadings, groups of links, and text are all separated. Con-sistency of design within your site will help. Draw attention to impor-tant elements with bold text, but do not underline anything that is not a hyperlink.

Most importantly, think from a visitor's perspective about how your site is laid out. If you were surfing the Web and came upon your site, would you find the most useful and dynamic elements first? Or would you have to wade through text and explanations in order to find the unique and useful elements of the site?

Remember that many visitors will spend merely seconds on your site without clicking to an interior page. If your pages are easy to skim and digest, then more will leave with a distinct impression of your site, even if they leave quickly.

Again, remember the importance of consistency, even if information on your site is updated or revamped often (as it should be.) If visitors to your site find information there that they want to see as it is updated, they are likely to bookmark the page. When they pull up the bookmark days, weeks, or months later, it's important that they find the page again.

Make Your Visitors Feel Confident

This is an essential element of good Web design. Because many visitors are not comfortable divulging personal information or conducting transactions via the Web, you must include elements that will put them at ease. As a Web-based business, you will have visitors and customers who are anonymous until they announce themselves. You want their e-mail addresses and other bits of personal information in order to better tailor your site to their tastes and in order to cultivate your client relationship.

Therefore, tell your visitors outright what your plans are with the data they share with you. Include a statement about what you are doing to safeguard their information. If you are not going to share the e-mail address they provide you with other vendors, then tell them so. In addition, give them the chance to say "no" to future marketing emails from your site—then respect their wishes.

Think about providing an entire page of information—a privacy policy—that divulges exactly how you will and will not use their information. Tell them how you accept orders, and the steps you take to ensure the security of your transactions. Tell them how you gather information and who has access to it. Tell them how you use the information you gather. Being up-front about information gathering and transaction policies will help customers to trust you with their information.

In addition to posting such a policy, consider posting customer testimonials on a page online. Ask satisfied customers to be your references and to provide you with such statements. In return, provide links to these customers' personal pages or make other concessions you feel are reasonable for their efforts. (Paying them for positive testimonials would immediately call into question the validity of those statements.)

Use Your Resources: Professional Design Agencies

Design agencies can add value to your site via its appearance alone. If you do not have a team of creative individuals who are savvy or who have the time to learn about Web design, consider seeking the help of a design agency. When choosing an agency, keep in mind pricing, experience, and flexibility.

If you sell a product in a retail shop or by catalog, you tell the customer the price up front. Perhaps you use a retail price and a sale price as important assets in your marketing scheme. Demand that the design agencies you consider give you the same up-front pricing. Ask whether they charge by the hour, by the job, for a full site of services, or by another formula. When you're starting a Web company, budgeting is essential, and Web design is seldom cheap. Companies charge anywhere between $500 and hundreds of thousands of dollars to build Web sites. Some services will be billed up-front, and some will not be included in your design quote. Make sure you know what services are and are not included. Hosting and secure servers, merchant accounts, software, and promotion services rarely are included in the up-front design charge.

Once you have some ideas about your site, write a concise letter for potential Web site designers. They can't give you a price unless they have some idea about what you want. If you can only afford $500 to design your site, then keep your letter short and concise. Companies won't allocate the resources necessary to thoroughly review a twenty-page letter for a $500 job. On the other hand, if you plan to pay the design company thousands of dollars for their services, then you may want to make the letter more formal and call it a Request for Proposal (RFP).

Whether formal or informal, the letter should describe your business, image, and target market; briefly list the important elements that you would like your site to include, and list URLs of sites you appreciate or that are similar to the design you want.

After you have written the letter, keep the following steps in mind:

1. Present your letter, paying special attention to defining the customers you want to target, to several Web site designers and ask for their feedback. See how their initial reaction and vision works with yours.

2. Get bids from several Web site designers before committing to one.

3. After you find the best candidate, put your design and payment agreement in writing.

Never hesitate to ask for references and examples of a company's work. Look for the elements of good design we have listed above. Are their sites professional? Do they all look alike, or is the company flexible with their design and able to use a company's unique selling point in order to create an original and effective Web presence? Does the design agency have e-commerce capabilities beyond design, such as designing payment systems and database systems for collecting and analyzing customer information? If not, another vendor may be able to offer you a more complete and cost-effective solution.

Finally, make sure that, after the Web site design is complete, you or someone on your team will be able to maintain and update the site. Know your ownership rights and what intellectual property is involved on your site. For instance, a Web site designer may use technology for your Web site that the design company uses for many Web sites. Other companies may own other programs used on your site, and some programs may have been designed specifically for your site. Make sure you understand what you will own, if anything, and how you may use the underlying technology in the future.

Your site may also include certain copyrighted and trademarked material, including original text, graphics and computer programs. If the designer is creating something for you to own, make sure he or she transfers ownership of this intellectual property in a written contract.

Table 4.1   Online Resources for Web Design

Name	URL	Description
Web Developers Virtual Library	http://www.wdvl.com/	Complete library of resources for both beginning and advanced Web designers
ZDNet Small Business	http://www.zdnet.com/smallbusiness/	Library of articles and other tips and advice on starting and maintaining a small e-commerce venture

Revisit Your Vision Statement and Goals

If you chose to enlist the assistance of a design agency, remember that you are asking for just that: *assistance.* You are responsible for marketing your services in a way that will turn visitors into customers. You must have a set plan and well-established vision for your company before you can fully utilize the services of a design company—or before you can fully utilize the services of your own partners in creating and effective Web design. After you have read about how you can best present yourself online, use your vision of an effective Web presence to reexamine your current business vision and goals. Hone or redefine this vision and these goals in light of your new knowledge. Remember that flexibility and innovation are a good portion of online success.

*Remember that flexibility
and innovation are a good portion
of online success.*

5

Keep Your Customers

Good online businesses today rely on customer relationships built upon mutual advantage and trust, the way businesses always have. Only today, the challenge is heightened by the global competition provided by the World Wide Web. Many products or services are likely to be available elsewhere at a competitive price. Your customer relationships will help define your practice and make your product unique.

Before you can sustain your success in e-commerce, you must ask yourself: How can I use the efficient technology available via the Internet and still create personal relationships? How am I going to maintain customer relationships in the digital world?

A warning: Don't make implementing good customer service and relationship building online overly complex. Although the technology to support some ideas for relationship building might be complex, the principles are simple. Your goal is to develop a great customer experience by giving the customers what they want: simplicity, service, and the ability to accomplish their goals without hassle.

In fact, although each of you reading will develop different answers to the previous questions, we guess that most answers will coincide somewhat with old business adages—those cornerstones of the mom-and-pop stores that still propel large chain stores today. Try "Service with a smile" or "Quality, Service, Cleanliness." Customers appreciate "The customer is always right." And never forget to say "Thank you" and "Please come again."

The trick is to translate these adages to online relationships. It's not difficult; it just takes planning in order to develop the appropriate support systems for customer service as you develop your e-commerce venture.

Follow these steps to keeping your customers:

1. Adapt "Service with a smile" to influence your responsiveness.
2. Redefine "Quality, Service, Cleanliness" for online content.
3. Use follow-up e-mails to say "Thank you. Please come again."
4. Use technology to find an ounce of prevention.
5. Be neighborly: Chat and discuss with your customers to cultivate your online relationship.
6. Suggest complementary products and services to increase both customer satisfaction and sales.
7. Keep your hands above the table. Let your customers know what to expect at all times.
8. Personalize your Web site for better service.
9. Let your customers control your relationships.

Service with a Smile

Adapting the principle of friendliness to service provided online might sound like an impossible task. After all, customers can't see you smile online. They can, however, judge from your eager reply that your company is glad to hear from them, appreciative of their presence, and open to additional comments. For example, at a party, the people who walk from group to group introducing themselves and talking to each group are considered friendly, polite, and outgoing. Those who keep to themselves are considered unfriendly and aloof. In this same manner, your quick attention to the customer and continuing online conversation with them will make your company seem friendly and welcoming of them as recurring patrons.

If you smile when you talk on the phone, customers can hear the difference in your voice. Without the audio cues of voice communication, it's much easier for customers to misunderstand you and your interest in their situation. Therefore, speed of response becomes essential in showing customers they are important.

When designing an e-commerce Web site, it should include some prominent means of contacting the company, such as a button for quick e-mails via the open Web browser. Customers may choose to contact the company with general product questions, or they may issue questions or complaints related to an online transaction. When the customer sends that e-mail, he or she will expect a quick reply. No more than twenty-four hours should pass between the e-mail and its response. A fast, friendly response will let your customers know that you value their opinions and online experience and that you are working diligently to keep them happy. As soon as the customer receives this response, a personal relationship, facilitated by the Web, has been established.

This very simple exercise in customer service makes a big impression on the anonymous Internet. However, too many companies lose their chance to create a personal relationship by being slow or not responding to e-mail messages. Having the time to sort, evaluate, and respond to each customer e-mail may feel like an overwhelming task to the entrepreneur consumed by other duties of the e-commerce start-up, but this chore should not be allowed to slip through the cracks.

When e-service vendor Brightware (www.brightware.com) tested the United States' 100 largest companies online, 15 percent answered the simple question, "What is your headquarters address?" within three hours. About 36 percent could not be e-mailed from their Web address. Another 10 percent never responded with the answer to the question, while some took up to twenty-three days to respond. Imagine that the question had been from a consumer who received a defective product and needed the address to return it. The customer, if he or she received a belated reply, would have felt undervalued in the transaction and would have been unlikely to return to that site.

To avoid this pitfall from the beginning, create a twenty-four-hour e-mail response policy that is respected companywide. Try designating one person as a sort of e-mail receptionist—someone who can monitor the general e-mail account and respond to general questions immediately, then sort and distribute e-mails to the individuals who can most quickly and effectively address the customers' more in-depth questions or concerns. Remember to designate a back-up receptionist for when the first person is away. Decide who will authorize replies and within what time frame, if authorization is necessary.

If responding to e-mails immediately will not be possible for a few days, or if a more in-depth response will take more than twenty-four

Hello! Thank you for visiting us at *www.website.com.* Please take a moment to give us your honest feedback, and we'll happily answer any questions you may have.

E-mail address: [   ]

How did you hear about Website.com?

❏ Online advertising

❏ Word-of-mouth

❏ Newspaper/magazine

❏ Radio/TV

❏ Other

Have you ever bought from website.com before?

❏ Yes

❏ No

When is the next time you anticipate purchasing a website.com or similar product?

❏ Never

❏ < 3 months

❏ 3–6 months

❏ 6 years–1 year

❏ > 1 year

Additional comments, questions or concerns:

[   ]

Thank you for your time. We at website.com should respond to your questions or comments within one business day.

Figure 5.1 Sample online survey.

hours to conjure up, consider using a free auto responder service available online (see Table 5.1). Companies can consider having these auto responders acknowledge all queries and designate how long they can expect to wait for a reply. Smaller companies can improve efficiency by developing standard responses to frequently asked questions and workflow processes that notify senior managers if a response has not been sent within the agreed-upon time frame.

Larger companies should consider e-mail response software agents (see Table 5.1) that can significantly automate the e-mail response process. A more substantial online customer service department could be created to handle nothing but customer correspondence. Amazon.com is reported to have more than 200 customer service representatives who handle customer e-mails, and as customers' complaints pop up on representatives' screens, they quickly assemble responses from an electronic library of more than 1,400 routine remarks. Amazon offers special rewards to those customer service representatives who handle messy situations with finesse and ensure a repeat customer. Representatives have the authority to quickly waive small shipping charges and other inconsequential fees that do very little to improve Amazon.com's bottom line but can alter a customer's perception of the company permanently and for the better.

Regardless of the size at which you start, remember that e-mail responsiveness is often your first and most efficient porthole for entering into personal relationships with your customers. It's one key to providing that one-to-one touch that will reward all of your online efforts with customer loyalty.

Table 5.1   Responding to E-Mails

Programs that allow you to personalize email	www.digital-impact.com www.guesttrack.com
Free auto responder services	www.myreply.com www.freedback.com
E-mail response software agents	www.brightware.com www.mustang.com www.delanotech.com www.mcsdallas.com (Calypso) www.aditi.com (Talisma)

In addition to responding to e-mails, in order to deepen your company's relationship with a customer, it is important to learn more about his or her wants and needs. You've already made your e-mail address prominent and easy to find, together with your toll-free number. Now consider the use of forms that can gather small bits of required information from your Web site visitors and potential customers. See www.freedback.com for free online feedback forms, such as those that ask for an e-mail address, the length of time before the customer next seeks to purchase your type of product, and a few lines for the customer's personal comments about your site, if they have any to add. An example of such a form is shown in Figure 5.1. Your response is likely to drop if you require more detailed personal information up front, before potential customers submit a simple question, feedback form, or request. About 12 percent of U.S. companies in Brightware's survey required such personal information up front.

If you do want to request information more personal than an e-mail address, or if you want to increase the likelihood that a correct or current e-mail address will be provided, post your company's customer privacy statement in a prominent location. If the statement is long, include a prominent link to that privacy statement in the text below your feedback forms or e-mail address. No one likes to discover unsolicited bulk mailings in the inbox of their favorite personal e-mail account. No company wants the poor reputation of causing such inconvenience and annoyance, either!

If your online business partners with other businesses and shares customer information for marketing purposes, then consider creating a box to check or uncheck on your response form that states, "Yes! I want to receive periodic newsletters about similar product information and services" or "Please send useful product updates and news to my e-mail account."

Quality, Service, Cleanliness

Traditionally, customers discern the quality of a product by word-of-mouth referrals from family and friends, from the look and feel of the selling establishment, and from physical contact with the product itself. Service and cleanliness help them determine the quality of a product and of an overall shopping experience. Customers expect top-quality products to be sold by attentive, professional-looking merchants. They will pay more for a shirt at Nordstrom, a high-end department store famous for its attentive personnel, than they would pay for that same shirt at a store with shabbier surroundings and customer service provided only at the cash registers or a customer service booth. The same holds true online. The professional look of a Web site, as well as the tone and choice of wording on that site, can contribute to the visually "clean" feeling that helps establish trust. Prompt replies to online customer service queries add with this "clean" feeling to promote a perception of quality online.

MARKET SPOTLIGHT

Customer Service Q&A with Author Juanita Ellis

E-commerce entrepreneur:

I am currently in the process of setting up an online storefront to sell handmade jewelry. My product line is typically sold to department stores in the United States. I currently send a brochure of my products to several department stores, and the buyers order between five and twenty-five pieces of jewelry. Once I go online, I will be taking over the responsibility of selling and servicing my customers. What are some of the ways I can ensure that my customers receive the service they need without having to provide twenty-four-hour customer service phone support, which is not an economical choice for me at this time?

Juanita:

You are focusing on the right area! Customer service plays an even bigger role when introducing and selling products on the Web. The key is to understand how your customer buys products and the type of information they need to make a decision. Think about the department stores that are currently buying your products. What type of information have they requested? For example, what are some of the unique features of the products? What is the cost of the product if they buy more than a quantity of five? Is there an incentive to buy significantly more pieces? This type of information must be provided to help the customer make the purchasing decision. Most importantly, now that you are going online, customers will want to know information such as shipping costs, return policies, delivery dates, and special packaging. Easy access to this information can be critical in making the sale.

So how do you provide all of this information? Let's start out with some basic techniques that have been used by both small and large companies:

Product Information

For each product, provide a short and a long description. The short description includes product name, description, cost, shipping charges, and delivery time. The long description should include special features of the product and other information in which you think the customer will be interested during the purchase. The more information you provide, the more comfortable potential customers will feel in deciding to make the purchase.

Develop a Question-and-Answer Section

Initially, when you set up the site, think about the types of questions that your customers ask when purchasing your jewelry. You may even want to ask some of the salespeople currently selling your products in department stores. For example, customers may ask questions like, "What do I use to clean the product?" or "Is the jewelry water resistant?" Customers also will want to know about the return and/or exchange policy. In addition, include a section where customers can enter their own questions and you can provide responses via e-mail. These questions may eventually be incorporated into your online question-and-answer section.

E-mail Response

One of the best methods to provide personalized customer service online is to include a confirmation of their purchase via e-mail. The e-mail should include the expected delivery date, total cost of the purchase, shipping charges, and a short "thank you" note. Take this a step further by including a small survey on their shopping experience. Over time, the surveys will provide you with valuable information about your customers' experiences while shopping your site. In turn, this information will give you feedback on what is working well and not so well on your Web-based storefront. Your own customers will tell you what you need to do to beat the competition and to succeed.

A 1-800 Number

I know you do not want to provide a twenty-four-hour, seven-days-a-week customer service hotline. However, consumers in some situations want to speak to a sales representative and ask additional questions in person, and others refuse to make online payments but will call and make payments via the phone. At the minimum, allow your customers to leave a voice mail at any hour. You may also consider hiring an answering service that is responsible for taking phone calls when your office is closed. This will help ensure that you are not losing potential customers who want to purchase a product but do not have the answer they need immediately or do not want to put that credit card number online.

> *In servicing your customer,*
> *put out the red carpet.*

No matter which of these ideas you choose to implement, keep in mind that the next online jewelry store is one click away for your potential and returning customers. In servicing your customer, put out the red carpet. Good customer service will differentiate your Web site from the rest of the World Wide Web.

The Customer Is Always Right

Merchants who implement this saying may take a few losses with untruthful customers, but their reward from good customers is immeasurable loyalty and confidence that customers will enjoy—or at least approve of—every shopping experience they have with the merchant.

Ensuring that the customer feels "always right" begins with the before-mentioned prompt reply to e-mails, even if they are substantial complaints that must be evaluated within the company. Answer all complaints promptly—within twenty-four hours is, again, a good rule of thumb. Have someone designated to answer complaints or to route them to the appropriate person for maximum efficiency in replying. If the complaint is serious and a offer of resolution will take time to develop, then send an e-mail notifying the complainant that their complaint has been heard, a proposed resolution is forthcoming, and give them some timeline for receiving the expected reply.

Along these lines, sometimes twenty-four hours will seem too long to wait for an answer to a complaint. List your company's customer service phone number on your Web site for those customers' complaints that feel more like crises. If no one will be available after hours to listen to a complaint, provide voice-mail access, and ensure a reply within one business day. Sometimes allowing customers or clients to vent their frustration when they need to is enough to win their trust and respect and to keep them coming back.

Mistakes within your company inevitably will happen. Prepare to keep your customers despite the occasional blunder by offering incentives for your customers to try you again. In a restaurant, a maitre d's apology for poor food or poor service may include giving the dinner to the customer for free. Online, an incentive could be an electronic coupon to apply on a future purchase or a free gift or product sample. Choose to offer any incentive that shows a customer that the blunder was happenstance and uncommon and that your company is committed to their satisfaction. Actively seeking their forgiveness and approval via prompt communication and an extra incentive to return will reward you in the long run with customers who feel valued, heard, and appreciated on the vast and lonely Internet.

Thank You, Please Come Again

Those words are still the golden phrase of customer appreciation, spilling from the lips of cashiers and clerks in small shops and large chains everywhere. They're the last thing a customer hears before leaving a store, and they remind a customer that his or her business was valued, that the transaction was pleasant, and that he or she is welcome to return.

Because of the extreme cost-effectiveness of e-mails, it makes sense to use that medium and follow up all of your transactions with similar "Thank you, please come again" e-mails that remind the customer that his or her business was appreciated. In a store, a clerk or cashier would be able to tell from a customer's tone and appearance whether or not he or she was satisfied with the transaction. Because the online experience takes away this personal responsiveness, these "thank you" e-mails should contain a means for the customer to submit feedback on the quality of his or her transaction with you. Send the e-mail from an address that allows the customer to respond by hitting the "Reply" key on their e-mail program. Ask them a short series of easy-to-answer questions about the quality of their transaction, and you're likely to receive a reply. The more lengthy the questionnaire, the fewer people will feel they have time to complete it.

The Web site Bizrate.com uses promotional offers, raffles, and give-aways to convince customers to complete more significant, voluntary surveys about the transactions conducted with its member businesses. Complete one of their surveys to get an idea of good questions that, when answered, could benefit your future Web site development and transactions. Larger companies could consider evaluating Bizrate.com for a list of member businesses and information on how to use their product to evaluate their own Web transactions.

If your online operation is still too small to make large-scale customer service evaluation cost-effective, develop your own "thank you" templates and online surveys. Don't feel there's any one wrong or right question to ask or that sending only one "thank you" e-mail is necessary. At several points in the transaction, even an extremely busy customer could appreciate follow up e-mails. Good times to send out "thank you" and follow-up e-mails include:

1. *As soon as the order is received.* Use this e-mail to thank the customer heartily, verify online payment and approximate time before the order is shipped, and include a customer service phone number for any questions the customer may have.

2. *When the order is shipped.* Use this e-mail to thank the customer again, provide a shipping date and package tracking number, estimate when the order should arrive, and include a customer service phone number. Most commercial shipping services have

package tracking systems that will help your online company facilitate this service.

3. *A couple of days after the order should have arrived.* At this time, the customer should have had a chance to evaluate the product, if it arrived as scheduled. At this point, you can offer incentives for referring other customers and discount coupons for returning for subsequent purchases. You can include e-mail addresses and phone numbers that will help track the package if it hasn't arrived on schedule. You can also include e-mail addresses and customer service numbers for product questions and complaints. Most importantly, this is a good time to send a brief questionnaire about the quality of the transaction, including satisfaction with the product and customer service.

4. *On a regular basis, via new product newsletters.* Place your customers on a biweekly or monthly newsletter list that notifies them of product upgrades, product news, or new products. Continue to send electronic coupons and incentives for referrals and return business. Don't pester your customers by sending too-frequent reminders, and provide a means for them to request to be taken off the product newsletter list. Customers who feel pestered by your frequent e-mails after the sale are unlikely to return or to refer others. On the other hand, if you don't send newsletters, your customers are less likely to see your name and brand in the frequency required for easy recall and true brand recognition. Remember that every follow-up e-mail should reinforce the favorable feel of the prior transaction and should entice customers to return. This will help reinforce a positive customer relationship.

Also, make it easy for customers to complain. When someone complains, he or she is presenting you with an opportunity to collaborate in solving a problem. You can then build loyalty and referrals. Just make sure you have a good process for handling complaints with some agreed service levels so you can monitor your performance. If you find few customers are issuing complaints, don't assume no one is complaining. Assume that your system for hearing complaints is failing somehow. Retool your Web site so that customers can more easily access you with complaints and can feel confident that their complaints will be heard and responded to; and, if necessary, you could consider offering a prize or some reward for the best feedback or improvement suggestion. Any criticism an e-commerce venture receives is constructive, not detrimental, and can only increase that company's competitiveness, if the criticism is heeded, discussed, and acted upon.

An Ounce of Prevention

Along the same lines, why wait for your customers to complain? Because of package and order-tracking software, the merchant should know before the consumer if a problem arises with the order. A forward-thinking merchant can make sure these checks on the purchase, processing, and shipping process are in place and can intercept problems before they reach the consumer. For instance, if the purchase did not go through, begin immediately remedying the problem with the hopes of saving the transaction. Customer service specialists should identify where the purchase went wrong, such as the wrong expiration date or misspelled name entered for a credit card purchase. The customer can be e-mailed quickly about this problem, with the hopes that he or she will choose to provide the correct information and allow the transaction to continue. The chance of the transaction never happening increases as the amount of time that lapses after the initial online purchase decision increases.

In addition, although automated e-mails can be precious boons to efficiency and can help respond to customer concerns quickly for maximum effectiveness, sometimes a truly personal response is needed to save the transaction or to ensure repeat business from the customer. If some element of the transaction has failed, and the result is significantly delayed shipping or an inability to provide the product at all, the merchant should work proactively and notify the consumer immediately. Otherwise, by the time the consumer realizes that the transaction has gone awry and takes the time to contact the merchant, he or she is already dissatisfied and frustrated with the transaction. The customer will appreciate the fact that the vendor cared enough to intercept and solve the problem rather than wait for the customer to navigate the customer complaint process. This "ounce of prevention" can not only cure whatever ill has occurred within that particular transaction, it can also solidify a trust-based customer relationship for life.

Be Neighborly: Chat and Discuss with Your Customers

In a bricks-and-mortar shop, customer relationships are built through the customary, "Hi, how are you?" small talk. Although such idle chatter would look ridiculous in any online format, similar customer relationships can also be built and maintained by the use of chat rooms and discussion forums. A chat forum hosted by a well-known expert in a particular field can be a good way to drive new and repeat traffic to your Web site, provided it is properly marketed. For instance, advertising a local golf celebrity's live online chat session on perfecting putting in the newspaper's sports section could drive many new customers to your site. Advertising the chat solely on your Web site would only encourage returning customers and would not increase your new-customer traffic flow. Hosting virtual events and conferences also could prove useful for some online companies who wish to court a very specific type of business-to-business client. At www.webex.com, find a free service for the sharing of documents and presentations and a program that enables customers to jointly surf the World Wide Web.

Discussion groups can not only create and maintain relationships between you and your customers but also can help customers get in touch with one another. These can be either unmoderated, in which case they may have more credibility, or moderated if you prefer more control. Moderating the discussions ensures that off-putting conversations are intercepted and your company image is maintained, even though the customers are doing all the work to lead, encourage, and develop the discussion sections. Discussion groups can be self-hosted, or you can utilize the increasingly powerful services provided by the free communities at www.deja.com or clubs at www.yahoo.com. These latter services can be links from your home page and allow messages to be sent to members, new members to be personally invited, and a number of relevant forums created.

Although online communities created via e-commerce sites seem plentiful these days, their popularity is driven by this simple truth: Online communities can be extremely powerful marketing tools. Arthur Armstrong and John Hagel in their book *Net Gain* divide these communities into four categories:

1. *Fantasy:* People assume new personalities or create new environments, such as combining players from different sports teams to

Fantasy	Relationship
Interest	Transaction

Figure 5.2 Types of online committees.

form custom "fantasy teams" that compete in online leagues against other such teams.

2. *Relationship:* People who have shared life experiences, such as divorce, difficult children or life-transforming diseases, come together, mainly to support each other. Some fun-to-read examples that promote a product or service include the communities on dating and relationships at www.therulesbook.com and www.whatmenwant.com.

3. *Interest:* People with a common interest in a topic gather to learn and share about the topic.

4. *Transaction:* Where people buy and sell products and services.

If you are not going to create an online community because it is too expensive or inappropriate, then you may want to consider ferreting out existing communities that you can sponsor or where you can place advertisements or editorials in order to draw new customers to your site.

Would You Like Fries with That?

Another means of valuing your customer and adding value to his or her experience is to suggest products or services that complement products or services that interest them. Businessmen and women with little free time will appreciate the convenience of one-stop shopping. Examine the products you are offering online, then ask yourself: "What can I do to add value? Will another product or service make this more useful or manageable for my customers? Can I offer to send them news or product updates that will help them while furthering our relationship? Is there a product or service available for purchase that will help my customer accomplish his or her goals?"

Add value to his or her experience.

Amazon.com and CDNow, for instance, create relationships among the products they offer online and instantly suggest similar or complimentary products and information to customers accessing their information online. Search for a particular book on Amazon.com, and you'll find a list of the books most commonly purchased by people who also bought the book you were seeking. CDNow will suggest music by artists and in genres favored by those who also sought your musician.

When you can add value to your product with other products, not only are your chances of grossing more sales from each customer higher, the chance is higher that they will return to find new products or services that will fit their interest or meet their needs. The trick is to make sure this information is usable and accessible. Make sure the suggested products are within eyesight of the products the customer already is seeking or is buying. If additional information can be provided on another Web site, make sure that information is only a click away and that customers do not leave your Web site in order to access this information.

Keep Your Hands above the Table

In some cultures, it is inappropriate at dinner to have one hand on your dinner fork and the other below the table. Why? Because it is unclear who else's hand you could be holding with that hand beneath the table. Online customers feel the same. They want to know exactly what both of your hands are doing before they're willing to put enough trust in you to complete a purchase. They want to know up front what products and shipping will cost, how long they can expect to wait for your product to arrive, and what they can expect from you if the transaction somehow goes awry. Knowledge of what to expect takes the fear out of buying and thus encourages transactions.

For this reason, the following information should be evident on your Web site:

1. *In-depth product information.* A short "teaser" description should be listed first, in order both to pique clients' interest in a new product and to help them find any specific product they seek. A link to more in-depth information should be easily accessible from this teaser description, perhaps by clicking on a thumbnail photo of the product or by clicking on the product name.

2. *Shipping charges.* If shipping is free, advertise it! Make this a significant selling point. According to a survey by Cyber Dialogue, www.cyberdialogue.com, free shipping was a major factor in the decision-making process for 90 percent of online adults who made holiday gift purchases in 1999. On the other hand, if you choose to have your customers pay for shipping, make sure these charges are evident up front. If these charges are not calculated and disclosed until the check out process, the chances are greater that a customer will feel shocked about the added charges and

Figure 5.3 What customers need when online shopping.

abandon the online shopping cart at the last minute. If you must charge for shipping, also consider offering shipping discounts for purchasing additional items. Disclose this freely as an advertising point, as well as an important notice to your potential customers.

3. *Shipping times.* Let customers know how long it will take to process their order and get the package out the door. If the product usually ships within 24 hours, display that information directly under the product. If the product is temporarily out of stock, and shipping could take an indefinite amount of time, display that information instead. Even if the delay in shipping could dissuade some customers from placing an order on that product, they will trust your promises of service on other, similar products.

4. *Product return and/or exchange policies.* Because these policies may be lengthy, provide a prominent link to the page with this information. Be warned that a short return period, an exchange or store-credit-only policy, and especially an "all sales are final" policy are serious obstacles to winning customer trust and gaining sales online. Because the customer cannot examine the product firsthand before making a purchase, he or she must have the perception that the online business stands wholeheartedly behind its products and believes just as strongly that the customer will either be satisfied or will have an easy channel for remedy.

5. *The shopping basket.* The shopping basket should be displayed for one-click access on every page, and when clicked upon, it should quickly display all products currently earmarked for purchase. The customer should be able to easily add or delete items or to change quantity and quickly recalculate total product costs and shipping charges. If navigating around and altering the shopping cart is difficult, the chance of the shopper abandoning the cart before checkout is astronomical. As we mentioned earlier in the book, few customers who select items for purchase ever check out with those items. Most online transactions are abandoned before they can be completed.

Personalize for Better Service

Relationships can also be built effectively and efficiently, simply by getting people to repeatedly return to your Web site. To get people to keep coming back to your site, you must provide them with a reason. This can be relevant, frequently changing information. Better yet, it can be a personalized page. For instance, personalized weather data available at www.weather.com (The Weather Channel) makes the Web site a quick source for one-click weather data among its customers. Customers can choose which cities or zip codes they would prefer as their primary weather information area, then they can choose a list of alternate cities whose weather data will be available for viewing via a quick pull-down menu. Customers can choose what weather maps they want displayed, as well as other personalized details.

At www.bluefly.com, the online clothing outlet store, customers can customize their own catalogs and thus choose to view only those products suitable for them. They begin by filling out a lengthy, specific questionnaire that includes the customer's preferred clothing and accessory designers, as well as the customer's shoe size, hat size, and every size in between. They can change the catalog at any time, but only clothing and accessories in their particular size and preferred style will be displayed.

Effective personalization does not have to include customer-specific content, however. Personalization can include maintaining password-protected private areas, using cookies that allow the site to remember the user on repeat visits, having personalized start pages (similar to many portals), building databases that produce personalized content, and creating different entry pages for separate categories of users. Larger companies may want to invest in personalization software, such as that offered at www.firefly.com, www.broadvision.com and www.guesttrack.com.

Perhaps the most effective personalization is remembering clients' purchasing information, so that they can engage in "one-click shopping." Streamlining the checkout process encourages customers to come back to your site rather than to seek out a competitor. Web surfers are lazy online. They don't want to get out their credit cards and type in all sixteen little digits, their birth dates, their shipping and billing addresses, their work and home phone numbers, and their e-mail address. They want to click on a product, choose "purchase now," and have it arrive on their doorstep within a few days. Online is all about convenience. Otherwise, online shoppers would still be found in bricks-and-mortar stores.

As an added bonus, not only does personalization of this nature encourage return customers and more online purchases, tracking customers' purchasing information also helps you, as the merchant, to directly target product and service promotions likely to interest specific customers, based on their past buying habits.

Customers Now Control the Relationship

In the digital world, customers now exist who report they have developed a more significant relationship with an online company than they have with their neighborhood grocer, despite the fact that they have no human contact online. This is because they feel more in control of the relationship with the online company, where they can get the information they want when they want it, rather than relying on some under-trained, under-resourced human salesperson in an offline store. With new technology such as Web-based video phones spreading rapidly, soon even people who want human contact with an online organization will be able to achieve it.

Therefore, while human relationships can be neglected on the Internet, this pitfall can be avoided with a little effort. In fact, a clever entrepreneur with the right customer service tools can provide more thorough, accessible, and personal service online than he or she could in a bricks-and-mortar shop. The challenge is to view the Internet as a tool with which to enhance our customer relationships, rather than an obstacle to creating them.

6

Fulfillment Channels That Work

It's no secret that the Internet has revolutionized the way goods are sold to consumers. Consumers have been empowered to select their purchases from a vast array of competitive services and to base their purchasing decisions upon boundless product information also available online. E-tailers have gone far in presenting and selling goods online. It's the transition from online purchase to tactile delivery—the fulfillment process—that has many merchants stumped.

Christmas 1999 brought the fulfillment crisis into the media limelight. Customers moved online in droves, hoping to find the perfect gifts without the headache of notoriously imperfect holiday crowds at traditional shopping malls. Although the World Wide Web could offer them the boundless product options and promises of prompt shipping that shoppers expected, shoppers found that many Web businesses could not fulfill their offers and promises. Web businesses' fulfillment channels were vastly under-resourced for the spike in online holiday shopping.

E-tailers found themselves spending big bucks to save face with their customers. For instance, when a bug in the Toysrus.com distribution software caused the Web site to overestimate its ability to deliver Christmas gifts in time, the company issued $100 gift certificates to customers who demanded to cancel their orders. Customers whose canceled orders arrived anyway were allowed to keep them free of charge. Even Web retailers, including Amazon.com, generally respected for its top-notch customer service, found themselves crippled. Shoppers frustrated by high-traffic-induced Web-site slowdowns tried to call customer service for help and faced an average hold time of thirty minutes at the company's call center. As a result, many shoppers' faith in online shopping was eroded, and their perception of the brands with which they experience flawed transactions were permanently altered. Companies who failed during the holidays last year may not be easily forgiven this year.

Clearly, neglecting fulfillment creates disastrous results. Fulfillment, representing the end of the site transaction, is a crucial part of the customer experience. You don't want to lose customers who have come this far.

Neglecting fulfillment creates
disastrous results.

Begin to establish fail-proof fulfillment channels through the following steps:

1. Learn the obstacles to fulfillment today.
2. Know your customers' expectations.
3. Know your shipping options, including national and global fulfillment networks.
4. Tackle other fulfillment areas.
5. Engage new information technology that enables high-speed fulfillment.
6. Discover e-fulfillment solutions.
7. Evaluate whether supply-chain technology could benefit your enterprise.
8. Plan for spikes by developing capacity flexibility.

Why Does Fulfillment Seem So Difficult Today?

As online shopping rose into the commerce limelight, the focus stayed tightly on marketing and product offerings online. The big questions were who was going to shop online, when they were coming online,

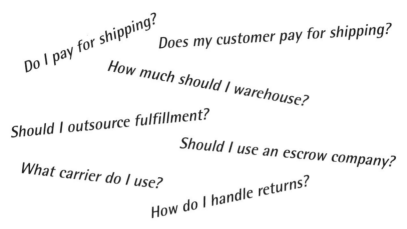

Figure 6.1 Considerations for the e-commerce entrepreneur.

and what they would be willing to purchase in a virtual store, without the tactile process of traditional product selection. Few questioned how they would get the products once they were purchased. After all, catalogs and mail-order businesses had been taking orders and filling them by mail for more than a century. How could filling an online order be any different? Why would fulfillment of online purchases require any special planning or concern?

The answer arrived loud and clear: Online transactions were different because they began in a virtual world. Few anticipated the challenges of moving from a virtual presentation and sale to the delivery of physical products. While shopping online, customers can obtain their desired products on demand, purchase them almost instantly, and have them delivered to the intended point of use. They expect the same instant efficiency and control over the fulfillment end of the process. Online merchants also change their online product offerings more fluidly and frequently than traditional catalog or mail-order merchants. An entire product line can change in a matter of minutes. As a result, both customers and the merchants themselves place untraditional expectations on what today remains a vastly traditional fulfillment process.

In addition, online fulfillment is still relatively new, and new technologies arrive daily to help facilitate the types of internal/external information exchanges and supply chains necessary for the fail-proof fulfillment of customer orders online. More importantly, these technologies help facilitate the exchange of information with the customer, in order to give the customer the same feeling of empowerment during the fulfillment process that he or she currently enjoys in the beginning of the transaction.

Know Your Customers' Expectations

Internet commerce has made it possible to buy and sell goods in near-real-time transactions. The customer is empowered to customize a good deal of the transaction process by choosing shopping, payment, and shipping options. However, customers do not expect the efficiency of the transaction to end with the sale. They expect the fulfillment process to proceed with the same visibility, efficiency, and dependability as their online purchasing experience. As a result, shoppers expect a greater level of efficiency from e-tailers than they do from traditional businesses.

At the same time, customers do not trust that this efficiency will be delivered. According to an Andersen Consulting study, fewer than half of online shoppers make a purchase online more than once a month. Although online transactions take a quick eleven to fourteen minutes on average, these shoppers still conduct the majority of their transactions in brick-and-mortar shops. Although an argument could be made that these shoppers are simply slow to break old habits, it's more reasonable to believe that some obstacle is keeping them from fully trusting their transactions to the Internet. These shoppers are obviously aware of the value and speed available on the Internet, because they do complete a dozen or more transactions online. Chances are that they hesitate because they doubt the ability of online businesses to meet or succeed their expectations for trustworthy performance.

The enormous amount of near-real-time information available online further heightens customers' expectations for online purchase fulfillment. When a customer has complete control over the shopping and purchasing process, he or she expects to be able to track the back end of the transaction with the same efficiency. A nervous customer who purchases a gift close to a deadline, such as a holiday or a family member's birthday, will want to know each step of the way how his or her order is being processed. Their inability to track this process physically may drive them to shop at a traditional store, where they know they will have a gift in hand as they exit the premises, even if the process takes substantially longer than an online gift selection and purchase.

If the customer is able to track the fulfillment process online, the merchant will face additional challenges. Many merchants today have leeway in the traditional shipping process that e-commerce merchants will not enjoy. A customer who placed an order with the L. L. Bean catalog five years ago would have found the middle insert that gave approximate shipping times and would have expected to wait. The process of getting that package to the customer's doorstep remained a mystery; all that mattered was that the product arrived within the promised time period. With the advent of online information systems and product tracking, online merchants will face tremendous pressure to perform accurately and on time at each step of the fulfillment process. Customers may want to know exactly what warehouse their order originates from, when their order is packaged, when their preferred shipper takes the package, and, finally, when they should plan to be home to receive the expected package. If this information is made available to placate eager online customers, then any misstep will be noted and will cause the customer to lose some faith in that merchant's capabilities.

Keep in mind that, right now, consumers seem willing to accept the shipping speeds that are offered by e-commerce companies. What they do not accept is promises broken. Therefore, promise shipping times and to meet holiday deadlines *only* if you have absolute certainty in your ability to do so. More important than giving your customers access to online tracking of shipments is giving them reason to trust your promises. If they engage in a flawless transaction with you once, their confidence will be buoyed. If they engage in a transaction where promises were broken, their trust may be permanently lost, and step-by-step tracking may be a necessity in order to regain their business.

Know Your Shipping Options

When thinking about shipping, first decide where your products will come from, then examine what fulfillment channels have a good reputation for the all-important on-time product delivery.

According to *The Standard* Internet business magazine, 43 percent of e-tailers ship merchandise ordered online directly from their brick-and-mortar stores' distribution centers. Another 23 percent ship the merchandise directly from these physical stores. Only 7 percent use Web distribution centers, and 18 percent outsource their Web fulfillment entirely. If you are moving online from a brick-and-mortar identity, then consider whether a Web-only distribution center is appropriate or whether your existing stores and distribution centers can handle the additional Web order traffic. If so, then using existing distribution centers could make economic sense, especially if your inventories are similar.

The reputation of your shipping resource will determine your own reputation for fulfillment, to a great extent. Right now, people trust FedEx and UPS to conduct the vast majority of e-commerce product fulfillment. They may not offer the most cost-effective solution for your company, however, so carefully researching your options is important.

Also remember:

1. Shipping is a two-way process in online retailing. Just as good fulfillment channels should facilitate the prompt and accurate delivery to an at-home address, the fulfillment channels must also facilitate easy product returns if you intend to have a strong return policy. Consider that you may pay for shipping the same product several times if you offer free shipping and the product is returned or exchanged by a customer.

2. Customers feel empowered when they can choose their preferred package carriers, delivery costs, and delivery times. They may trust the fulfillment process more if it is facilitated by a familiar shipping service, and they may be less likely to abandon their online shopping carts in favor of a brick-and-mortar store where they know they can receive the product that day.

3. Most shipping services track package delivery internally, while some may offer electronic interfaces that allow your company to monitor package delivery with equal speed and access.

4. Some shipping services offer interfaces that allow customers and merchants alike to evaluate shipping options easily and accurately. Popular sites include www.iship.com and www.smartship.com.

MARKET SPOTLIGHT

Shipping Q&A with Author Juanita Ellis

E-commerce entrepreneur:

What is the best method to ship products to customers buying from my Web site? Currently I am using FedEx as a primary method of shipment, but this seems to be quite expensive. My products range from coffee mugs to T-shirts. Any suggestions?

Juanita's reply:

Selecting the right carrier for the e-tailer has always been like navigating a maze. We are hit with so many options that the best choice is not always as clear as we would like it to be. The bottom line is that shipping rates are determined by three major factors: distance between origination and delivery points, package weight, and required speed of delivery. Let's say you want to ship a 5-lb package from Los Angeles to Baltimore. Using FedEx Priority Overnight would run approximately $35, while UPS Next Day would run approximately $33 for the same package. Both packages will arrive by 10:30 A.M. the next business day. Other carriers, such as Airborne and USPS, offer noon-deadline deliveries. Airborne charges about $23 for the same 5 lb package, and USPS charges $24.

There are several online sites that provide information on the best shipping options based on origin and destination of the package as well as the weight and dimensions. And guess what? *It's free!* Let's cover some of these services from the online vendors.

For example, www.iShip.com provides an excellent interface that gives you the best options for shipping a package based on delivery times, weight, and destination. iShip provides a comparison of UPS, FedEx, Airborne, U.S. Postal Service, and Yellow Freight. When going to the site, the customer is presented with an online form to complete with shipment dimensions, postal codes, and loss protection options. After the form is submitted, the Web site derives a matrix containing all of the shipping prices of UPS, FedEx, Airborne, U.S. Postal Service, and Yellow Freight.

iShip also has a tool for merchants to integrate the shipping comparison service directly into their Web site. With this tool, merchants can offer buyers the option to compare shipping methods online and then make their own selection from your site. This will give them accurate shipping estimates. When you, as the merchant, provide such accurate information, you don't have to worry about losing money on shipping, either.

Another company, www.SmartShip.com, provides very similar services, where you have the option of comparing shipping costs per carrier. Prior to shipping a package, you enter in the standard shipping information, and SmartShip will provide each carrier's pricing model based on destination, weight, and delivery dates. SmartShip also has a tool that may be integrated into your site so shipping prices are automatically updated and displayed based on the buyer's selected shipping method.

So, to answer your question, "Is FedEx the best-priced method for shipping? Is there a cheaper option available?": The two most common carriers used for shipping by e-businesses are FedEx and UPS. However, you may want to offer several options to your consumer, so that they may choose the carrier of their choice. They will feel empowered by the choice in pricing and delivery times.

Shipping and Your Bottom Line

As you consider which shipping options will offer your e-business the right combination of trustworthy shipping and cost effectiveness, also consider the following: What order size will make shipping with different carriers benefit your bottom line? A top-notch direct fulfillment company may expect to spend $9 to $10 per order on processing costs, while average performers may spend about $15 per order. You, as the retailer, must either charge your customers these significant shipping costs to recover your losses or somehow increase your average order size to recover these shipping costs. Your average order size may have to hit $50 or more in order to turn a profit. One health products online retailer offered, as a promotion, free shipping to customers placing online orders that totaled $40 or more. That merchant is offering an incentive for customers to help protect its gross margins.

Offering free shipping on at least some orders is a powerful incentive to your customers. According to a survey by Cyber Dialogue (www.cyberdialogue.com), free shipping was a major factor in the decision-making process for 90 percent of online adults who made holiday gift purchases in 1999. Forrester Research (www.forrester.com) confirmed this data in a separate report. Many online retailers sensed this and picked up the delivery tab for customers during the holidays. They tossed an appealing carrot in front of the 82 percent of consumers who say that the cost of shipping matters and would affect their decision to shop online. The result: New customers moved online in droves for the holiday season. If you are not offering and adamantly advertising free shipping as a service, you may want to think again.

Nordstrom, the popular upscale department store well-known for its superior off-line customer service, transferred this strength to its Web site with a flexible and easy-to-find return policy that is one of the clearest and most comprehensive on the Web. From the home page, customers can find the return policy by clicking a link labeled "Easy Returns, Free Exchanges." This takes them to a page entitled "5 great reasons to shop at nordstrom.com." The page tells customers how to make a return, explains the return policy, and provides other information. Customers see that they have two options for returns. One is to place their purchase in a preaddressed envelope for the mail carrier to pick up. The other is to return items to a brick-and-mortar Nordstrom—an option that shoppers who live near a Nordstrom would appreciate. This hassle-free return policy is extremely popular and thus serves both Nordstrom and its customers well.

Can Not *Shipping Be an Option?*

It may make sense to restrict where you will ship your product if delivery to some foreign or domestic locations isn't enabled by either your budget or your current transaction and fulfillment systems. If you choose to ship only within certain areas or countries, remember that any Web presence is, by nature, global. Don't let your customers fill an entire shopping cart and attempt to check out before they receive notification of shipping restrictions online. Have your shipping destinations listed as prominently as your delivery methods and times to avoid customer frustration.

Shipping also may not be the best choice even when it's entirely feasible to ship your product. For instance, if you're expanding your brick-and-mortar retail practice to the Internet, consider that your best fulfillment option in rush situations may be to create an incentive for an online customer to come into the store. The customer who wants a dress or shirt for a party that night or a band's CD before an imminent concert may be willing to use your online store as an important existing distribution channel. One solution may be to mesh the inventory tracking

Table 6.1 Common Carriers

The following popular carriers are all included in www.iShip.com, which is now part of Stamps.com, and www.SmartShip.com. Customers can easily weigh shipping costs and options via both Web sites.

Federal Express	www.fedex.com
United Parcel Service	www.ups.com
Airborne Express	www.airborne.com
United States Postal Service	www.usps.gov

These smaller carriers also are included at www.SmartShip.com:

UltraEX	www.ultraex.com	Offers national same-day delivery within a fixed number of miles from package origin and many overnight options
AirNet Express	www.airnet.com	Provides national overnight services with late-night pickup Monday through Thursday
Golden State Overnight	www.goldenstateovernight.com	Provides afternoon pickup and next-morning delivery to any address in California and Western Nevada

For same-day, e-commerce fulfillment:

| NextJet, Inc. | www.nextjet.com | Offers same-day, next-jet-out delivery from doorstep to doorstep and technology that allows merchants' online customers to arrange shipping from their point of purchase |

systems of your brick-and-mortar and online stores and then allow consumers to discern online whether the item would be in stock at a store near them. In other words, fulfillment can come in many forms, and your Web site can be more than a means of getting products shipped to the home. It can also be a brand presence and important informational tool.

What Are My Other Business-to-Business Shipping Options?

If the common carriers listed in Table 6.1 cannot provide good options for shipping your products in bulk to other businesses, a quick search via an Internet search engine will reveal a virtual smorgasbord of vendors waiting to help you fulfill your online orders. Searching www. yahoo.com for "shipping" produces a list including 791 cargo services, 67 courier services, 180 postal services, as well as categories for product-specific transport services, such as the transport of animals and vehicles. Many large business-to-business shipping companies also offer supply-chain management for meeting demand, which we address later in this chapter.

One major B2B package carrier is BAX Global (www.baxglobal. com), a $1.8 billion global transportation and supply chain management company that offers freight forwarding to business-to-business shippers worldwide. The company has its headquarters in Irvine, California, operates more than 500 offices in 121 countries, and employs more than 7,600 people.

C. H. Robinson (www.chrobinson.com) is a $2.3 billion transportation and supply-chain management company that has its headquarters in North America and has 3,314 employees working among 132 offices worldwide.

BAX Global (www.baxworld.com, formerly Pittston BAX Group) ships to more than 120 countries and helps customers keep track of their loads. Overseas corporations account for about two-thirds of the freight transport and logistics company's revenue. In the United States, the company offers door-to-door delivery and overnight service between major cities. BAX Global ships by ocean and air, acts as a broker to clear items through customs, and maintains warehouses and distribution points worldwide.

Another carrier, APL (www.apl.com), provides worldwide container transportation by sea via its fleet of 76 ships and 450,000 deployed containers. Subsidiary ACS Logistics provides supply-chain information systems and cargo-consolidation services.

Emery Worldwide (www.emeryworldwide.com) provides global air and ocean freight transportation logistics management, customs brokerage, and expedited services to manufacturing, industrial, retail, and government customers. Emery specializes in international shipments to 229 countries through a network of more than 600 service centers and agent locations around the world, door-to-door transportation to and from any location in North America, and business-to-business moves for heavyweight shipments of 5 lb or greater.

Tackle Other E-Fulfillment Issues

Other important e-fulfillment issues often neglected in favor of shipping issues include online escrow and business credit rating companies. In large business-to-business transactions, employing both types of fulfillment channels can ensure smooth and profitable transactions, as well as erase the doubts businesses with no prior relationship have about entering into transactions with one another.

Escrows

Escrow companies are impartial services that hold the money during online transactions and monitor the transaction to ensure that the deal is completed on time and to the satisfaction of all involved parties (see Table 6.2). The goal is security for both buyers and sellers. Sellers want to know that the buyer's payment is received before the goods are shipped, and buyers want to know that their funds will be released from escrow only after they approve the transacted merchandise. The method of payment, be it cash or credit, is often negotiable.

Table 6.2   Common Online Escrow Companies

Escrow Net International	www.escrownet.com.au
Escrow.com	www.escrow.com
i-Escrow, Inc.	www.iescrow.com
TradeSafe	www.tradesafe.com

Credit Rating Companies

Credit rating companies provide debt verification and general up-to-date business credit ratings for the purpose of entering assuredly into business-to-business transactions. Many credit rating companies are mostly country specific, such as the United Kingdom's CreditStatus.com or India's www.careratings.com.

In the United States, a recognized leader is Dun & Bradstreet (www.dnb.com), which offers a broad range of global credit investigation and reporting solutions for corporations of all sizes to minimize risk and maximize revenue. D&B credit products and services are designed to offer up-to-the-minute credit information, in-depth reports, and exclusive risk indicators and scores. They include software solutions for automating the credit decision-making process, as well. Available reports include business background and information reports, European credit reports, international business information reports, and public records searches.

As with any e-commerce solution, each company has its own particular strengths and weaknesses, with many company databases covering a particular niche not included in a competitor's data. One corporation that claims to cross these boundaries is Instant Search (www.instant-search.com). The site was developed for those who want access to commercial business information databases but did not want to pay a subscription for occasional usage. Instant Search has links with all the commercial credit reference agencies and marketing databases, plus in-house access to a large number of libraries in both electronic and, in some cases, hard-copy format. The company has its own electronic data banks and links to international sources of company information.

Engage New Technology

The key to flawless fulfillment is heightened information exchange: the integration of order management, inventory management, capacity management, customer service, and delivery. Integrating such data would be an overwhelming, perhaps impossible, task without the advent of technology to facilitate real-time information sharing.

Traditionally, proprietary electronic data interchange (EDI) systems were developed internally by businesses to fill their needs. An Internet standard now exists for data interchange, called XML. This standard allows trading partners to give data certain tags that make it meaningful and useful in their particular information exchange.

Yet most systems integration today still relies on small-batch EDI and the copying of files. The result is sluggish fulfillment and bloated inventory. These traditional systems prohibit fluidly changing trading partners or business models. They prevent e-commerce fulfillment networks from moving at the speed of the Internet.

New information technology architecture integrates external systems with internal systems. More architectures will be planned to accommodate external data and to be deployed across multiple trading partners, while allowing each trading partner to customize the internal "look" of the data to best suit their processing needs. The goal will be real-time information sharing across multiple networks, which will facilitate e-fulfillment networks' inclusion of visibility and will allow the real-time tracking of shipments by all partners, customers, and authorized parties. These networks also will enable visibility to reduce stagnant inventory, to streamline distribution, and to improve performance internally and with trading partners.

Find E-Fulfillment Solutions

More and more companies are beginning to offer solutions for outsourcing complete e-fulfillment solutions. One of the newest is ebHub (www.ebhub.net), which is designed to provide midsized merchants with complete and comprehensive services to help their businesses grow. Unlike traditional ways of building an online presence, ebHub offers merchants one point of entry where they have access to services, including first-step site launch capabilities and, in the end, selection of their shipping company. ebHub is the first to help furnish payment processing, shipping, fulfillment, customer care, and marketing services through a fully integrated network of globally recognized service providers. ebHub promises that its merchants will have maximum scalability as their businesses expand, be able to improve the quality of their online offerings, and ensure profitability of their storefronts.

Also popular is Descartes DeliveryNet solutions (www.descartes. com). DeliveryNet includes industry-specific e-fulfillment software and Web-hosted applications that allow the seamless integration of business processes across a fulfillment network.

Click Logistics is focused on reducing your transportation costs, increasing your service levels, and automating and simplifying the shipping process. They have hundreds of carriers and manage tens of thousands of transactions for shippers like you. Logistics Management services include:

- Transaction management
- Mode analysis (parcel, LTL, TL, intermodal, air, ocean, etc.)
- Consolidation/deconsolidation
- Optimization (rate, mode, carrier)
- Continuous move management
- Carrier management
- Information management
- Claims management

VerticalNet's Logistics Online (www.logisticsonline.com) community offers additional links to trade partners, its logistics marketplace, and information about transportation and other fulfillment issues. As a whole, VerticalNet (www.verticalnet.com) offers a myriad of business-to-business e-fulfillment solutions through its industry-specific online communities. VerticalNet's offerings for building and managing centers of business include:

- Targeted communities where professionals and companies in a particular industry find fast, efficient business information, interaction, and transaction opportunities.
- Targeted content, including wide-ranging product and supplier information.
- Targeted solutions for increasing sales and reducing costs through effective online business connections and e-commerce avenues.

VerticalNet helps online buyers and suppliers worldwide to exchange information, source products, and execute online transactions. The Internet can be a powerful tool for building bottom-line growth, and Vertical-Net helps companies to achieve their e-commerce potential with solutions that effectively increase speed, efficiency, and their return on investment. In addition, VerticalNet has developed unique technologies that allow buyers to source and compare complex products online and suppliers to participate in e-commerce without developing a costly infrastructure.

E-Fulfillment Supply-Chain Management

We've already established that customers are growing increasingly impatient with fulfillment systems that cannot keep up with the speed of information and transactions on the Internet. For competitors on the Internet, this means that the winners will be able to optimize the flow of goods and services so customers receive a unique and positive fulfillment experience.

Many industry participants believe Internet-based supply chain management will revolutionize the way businesses bring products to market. Virtual supply chain integration networks will optimize established relationships of supply chain participants by streamlining the flow of information between all participants. They will connect numerous work environments and facilitate the real-time exchange of communication data. Cahners In-Stat Group, an e-business research group, predicts that these networks will eventually replace today's electronic marketplace structure, will lead to the eventual integration of marketplaces and online supply chains, and will result in an unprecedented level of connectivity among participants.

Good Internet supply-chain management systems should be able to suit the three Ts: time, transparency, and trust. In addition, they should:

- Improve responsiveness to supply and demand fluctuations.
- Minimize inventory without sacrificing customer service.
- Improve reliability of delivery commitments.
- Maintain profit margin.
- Help establish global transportation networks.
- Synchronize manufacturers, suppliers, and logistics providers.
- Accommodate changes in product inventory, business practices, and trading partners.
- Create systems to leverage technology and processes among many customers, creating network effects such as group transportation buying power.

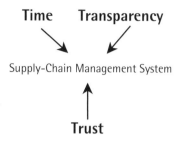

Figure 6.2 Factors for good supply-chain management systems.

- Preserve the diversity of trading partners' business processes and information systems.

We should note that some commentators have declared supply-chain management "dead" in this Internet age and have noted that general efficiency in factory operations have tossed aside any necessity for the field. Stanford University's Global Supply Chain Management Forum reports the following on its Web site, www.stanford.edu/group/ scforum/:

> Due to the recent trends of vertical disintegration, international procurements, new information technologies and increasing pressure from customers on responsiveness and reliability, and the globalization of operations and markets, supply chain management has become at once a challenge and an opportunity. Indeed, many companies have now the viewed supply chain management as the core of their business strategy.

The Forum brings faculty and students from different schools, departments, and disciplines together to work on theoretical and model based research, empirical research, and detailed field-based studies of supply-chain problems.

We share in The Forum's belief that automating the supply chain is one of the loftiest goals that a business can attempt to reach. The ideal is to create a business system that functions automatically wherever possible and reasonable. More typically, however, companies are content to improve their processes incrementally, automating where practical. Because the strength and coherence of their relationships to suppliers and buyers define most businesses, improvements in supply-change management can greatly improve a company's productivity.

The benefits derived from effective supply-chain management compel companies to contribute even more resources to the management effort. Such benefits include reduced expenses. Less time is spent on pushing paperwork, phone calls, faxes, and on tracking all of this information. For example, the typical purchase order costs between $75 and $125 to process manually. With the improved automation that e-commerce offers, that cost can be reduced to about $3. Bills can be customized for each client. Because many of a company's processes are automated, errors are naturally reduced (and often eliminated). The concept here involves creating a process that performs correctly and repeating that process. On the inventory management side, an online effort can result in just-in-time inventory, reducing storage and handling costs, and, if you manage clients' inventory, you can lock out the competition. All of these improvements are available within the supply chain, without even having to address the sales side of the equation.

i2 Technologies (ww.i2.com) leads the industry in supply-chain technology. i2 offers RHYTHM Supply Chain Management (SCM), which integrates all subprocesses that enable the exchange of information and the movement of goods between suppliers and customers, including manufacturers, distributors, and retailers. These subprocesses are demand planning, supply planning, and demand fulfillment. Often, customers' end goal in using the solution is to increase overall supply chain agility at the lowest possible cost.

Another strategy for improving supply-chain management involves implementing an XML-based back-office system for applications such as supply-chain management, automated procurement, shipping and logistics, and aggregation of business intelligence. XML (eXtensible Markup Language), the universal language for structured data on the Web, can drive cost-efficient, reliable, and open means of communicating between business partners and customers. Companies can leverage existing investments in virtual channel sales, marketing, customer service, and purchasing systems by replacing current and proposed electronic data interface (EDI) systems with XML. Customers learn that XML is easier to deploy, is easier to understand, and can be adopted much more rapidly than traditional EDI solutions.

webMethods (www.webmethods.com), a company focused on delivering XML-driven products and services, enables enterprises to forge automated links with customers, suppliers, and partners quickly, easily, and cost effectively. Their B2B products enable companies to leverage investments in existing Web sites, applications, and local data sources, reducing the time and complexity of implementing true business-to-business e-commerce compared to technologies such as CORBA and traditional EDI.

The resulting benefits of this XML deployment are compelling. Clients gain tight intercompany integration; linked Web sites, ERP systems, databases, and EDI systems; improved customer service; leveraged current IT investments; an extensible, scalable solution path for their business partners (with no purchasing required); and a low cost of deployment. Implementation of an XML-based solution can be accomplished in weeks, instead of months or years.

By allowing companies to reduce the expense associated with their supply chain, some facilitating companies exploit the Internet very successfully. For example, the National Transportation Exchange (NTE) at www.nte.net, a Chicago-area company, has created an Internet-based exchange for empty truck space, promising to slash $15 to $20 billion a year in distribution and transaction costs from the $400 billion trucking industry. The company's president, Greg Rocque, and his staff have enlisted 350 members so far, and expected to handle approximately 150,000 loads during 1999. Members pay a startup fee that varies according to the amount of deployment work that NTE engineers must perform to mesh a member's databases with NTE's. NTE also collects a small fee for every match they make.

Plan for Spikes

Even if you choose to engage supply-chain management solutions that help predict supply and demand, you still must plan and prepare for temporary spikes in demand. As e-tailers found during the 1999 holiday season, not planning for these spikes can cripple consumer confidence while it cripples your fulfillment capabilities. Spikes can be dangerous because they push capacity to the limit, and, so far, e-tailers are finding that postspike traffic doesn't settle to its previous level. Retailers actions this spring and summer demonstrate that they have learned from last year's experience; they are beefing up their inventories and leveraging their brick-and-mortar assets. Gone are any last-minute plans to revamp sites or to cut corners on distribution or fulfillment.

The simplest measures to plan for spikes without the help of software solutions include:

1. Research newsletters, user groups, and other Web-based information channels to discern what your biggest-selling products may be in any particularly busy time, such as the holiday season. Collect as much information as you can on what consumers want, and plan ahead to secure appropriate supply. This is a risky move but could benefit your bottom line if your research is thorough and your intuition is on target.

2. Finish any Web site or business process alterations two or three months before an anticipated rush to avoid site outages or other major infrastructure-related problems during the crunch.

3. Plan for securing temporary workers, if needed. Because competition for temporary workers can be fierce, begin relationships with local temp agencies long before you anticipate requiring their services. This should prevent shortages in customer care fulfillment, such as the long hold times the Amazon.com customer service line saw during the 1999 holiday crunch.

4. Work with your shipping fulfillment channels to make sure your shipping processes are easily scalable to significant increases in traffic. Forewarn them of your expected period of increased shipping, and remember that traffic may remain elevated after the spike has passed.

Williams-Sonoma, the popular furniture and housewares store, decided to address customers' special-occasion delivery doubts straightforwardly on its Web site. The merchant's site clearly tells customers when orders must be placed in order to arrive on time. This information alleviates customer worry and prevents disappointment over late deliveries. When customers click "Shop" on the home page, they are taken to the main product page that presents various product categories. In the bottom right-hand corner of the product page, Williams-Sonoma informs customers: "Order by 3:00 PM (PST) February 9th for Valentine's Day delivery." If customers click on the text, they go to a page with order deadlines for Valentine's Day and other forthcoming holidays, such as

Passover and Easter. The page further states that customers who miss the order deadline by a few days can rush the delivery for an extra $10.

Remember that the best planning for spikes mirrors the best planning for dependable e-fulfillment at any time. This best planning happens when you employ the information technology and software solutions that can optimize the fulfillment cycle, increase fulfillment visibility, and thus enable fulfillment systems to quickly handle increased capacities whenever necessary by synchronizing trade, supplier, and shipping partners' diverse systems.

7

Choose Your Payment Systems

When buyers and sellers engage in commerce, they agree upon what is being sold and at what price. The logical outcome is systems of payment, as sellers deliver the goods as promised in exchange for compensation. In the virtual world, screens display images of goods, and mouse clicks on "Buy Now" buttons reflect buyers' acceptance of purchases. FedEx or FTP typically replaces the sellers in delivering the hard goods or digital content. But the core commerce transaction remains the same and requires the secure exchange of payment. This commonly has become the first technical and psychological obstacle for most merchants' e-commerce efforts.

Separated into discrete parts, a transaction is comprised of an offer, authentication, payment, and delivery. An e-commerce site and associated systems must protect the interests of both the buyers and the sellers during each part by providing security and integrity. Security protects the seller from illicit use of the payment instruments while shielding the buyer from misrepresentation. Integrity guards the buyer from an unauthorized disclosure of the transaction and shields the seller from disputes about the timing or terms of a sale. These requirements significantly raise the bar of complexity for Web sites and protocols.

Electronic payments generally are representative of real-world exchanges. Payment protocols provide an electronic counterpart to paper-based methods of exchanging goods. In fact, if performed correctly, electronic purchases prove far more secure than other purchases, as traditional, paper-based methods frequently display complete and unencrypted credit card numbers.

Most credit card transactions within the U.S. economy occur over dial-up lines between merchants and cardholders. In the Internet economy, many sites still rely on these protocols; this is especially true for merchants with existing retail systems in which a Web site acts as a counterpart to brick-and-mortar storefronts.

Electronic payment systems handle the monetary exchange transactions for goods and services. Along with traditional monetary instruments such as bank notes, drafts, and credit cards, the e-commerce world employs a variety of tools. However, each tool can be judged on its conformance to the following requirements: independence, security, privacy, anonymity, transferability, divisibility, ease of use, and total cost.

Follow these steps in order to choose payment systems that will best facilitate your transactions:

1. Learn the various categories of electronic payment systems, including new smart card options.
2. Find the category that dominates e-commerce.
3. Examine the differences in how transactions work in the physical world versus online.
4. Plan to put the components to enable electronic payments into place.
5. Evaluate payment methods for their ease-of-use by all parties involved, including consumers, merchants, and financial institutions.
6. Know the actual costs of a transaction.
7. Set up credit card verification.
8. Enact protocols that ensure online security.
9. Help your customers develop trust in online transactions.

Payment System Categories

Electronic payment systems fall into a variety of categories, based on their function. Micropayments, although measured in pennies, will enhance several aspects of e-commerce when a more innocuous busi-

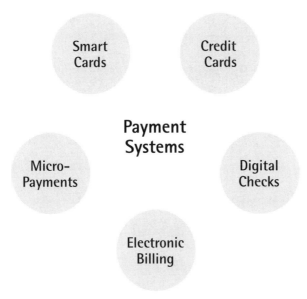

Figure 7.1 Different forms of payment for online purchase.

ness model is implemented. Smart cards, either disposable or reloadable, store monetary value on a card that can be debited. Electronic billing enables the presentment, payment, and posting of bills on the Internet. Open exchange relies on traditional bank or credit card clearing procedures to process a transaction, typically initiated by an e-mail message or HTTP form. Secured linkage is the encrypted transmission between merchants and customers where security-enhanced software and front-end browsers provide authentication. A trusted third party, as a mediator, handles all credit or check clearing functions between merchants and customers. After the process is complete, the trusted party provides the transfer of funds. Digital check, another category of electronic payment systems, replaces the traditional check with the security of an encrypted transmission. Digital cash resembles cash with its serial numbers, but it travels electronically.

These categories represent generalizations about the function of electronic payment systems. As a practical matter, your customers are likely to compensate you for goods or services in one of four ways: micropayment, smart card, electronic billing, or credit card.

Micropayments

Imagine you want to read an article from a subscription magazine published on the Web. Instead of paying for a year of subscription service or even a single issue, you might be willing to pay for only the specific article. Micropayment technology allows consumers to pay for digital content in increments of cents or fractions of cents. Such content typically includes music, videos, published articles, or software. The advantage of micropayments is that funds do not have to be reloaded for each transaction, and minimum transaction amounts are much lower than for credit cards.

A typical micropayment transaction consists of four steps:

1. The customer sends a request to load funds using his or her wallet software.
2. Via the banking network, the funds are verified as available in the customer's bank or credit card account. Bank accounts and credit cards are handled differently as sources of funds.
3. After approval, credit card funds are available immediately, while bank account funds might require up to a five-day wait.
4. After the money is transferred into the wallet, the customer can make micropayments at Web-based stores. The wallet automatically debits the amount of purchases from the balance. One of the main advantages of the micropayment method is that funds do not have to be transferred from a credit card or bank account each time a purchase is made.

The challenges of reasonably implementing micropayment systems are twofold: delivery of the technology and charging for digital content. Several companies have established reliable business models based on micropayment technology, particularly on the delivery side. Qpass sells content from publishers such as *The Wall Street Journal Interactive Edition* on a short-term or per-article basis. Cybergold allows users to purchase digital content such as MP3 songs, software, and video files. Both companies have set up micropayment programs to address the challenges of charging small amounts online. These companies receive a portion of their revenue as fees from content partners.

One disadvantage to the micropayment system is apparent from the customer's perspective. Typically a customer might not object to paying

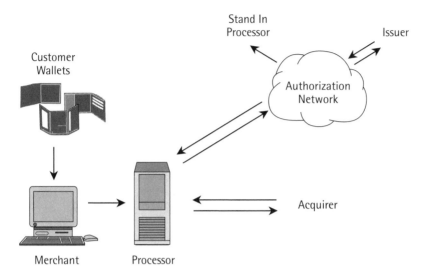

Figure 7.2 Micropayment transaction.

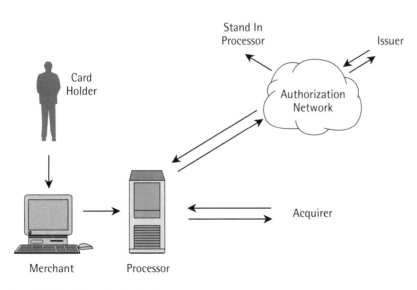

Figure 7.3 Credit card authorization.

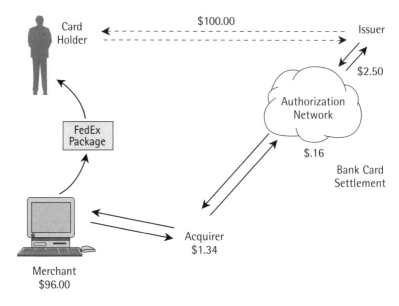

Figure 7.4 Credit card settlement.

five or ten cents to read an article or to access some specific information. However, multiply this experience twenty times within a single Web session, and your customer becomes exasperated. The merchant, in this example a publisher, would increase the revenue stream by charging a per issue, quarterly, or yearly subscription.

There are some micropayment systems in use online. The two micropayment models are debit and credit. A debit system like Compaq's MilliCent™ Microcommerce Network (www.millicent.com) most closely mirrors a cash payment system. At the beginning of the month, buyers charge a certain amount to their micropayment accounts, and MilliCent deducts each payment from their electronic wallet. The disadvantage of MilliCent is that few online stores accept it as payment, making buyers' cash deposited into the account nearly useless.

Microsoft Wallet, which we also address as an electronic wallet payment method, is a credit-model micropayment system built into Internet Explorer 4 and above. Microsoft Wallet collects all the charges you make, pays them, and sends you one bill at the end of the month. The disadvantage is this system's tracking of purchases, which more mirrors credit card payments than cash payments. The advantage is that Microsoft has the muscle to get vendors to support its payment program. The program is most popular in Australia.

Smart Cards

The smart card is an integrated circuit chip embedded in a small piece of plastic. Some models of these chips can hold 100 times the information contained on a standard magnetic-stripe card. The chips also add "smarts" to the card, allowing it to store and to process information. Smart cards are available in disposable (or memory) and reloadable (or processor) versions. Disposable cards have value the user can spend. In contrast, most reloadable smart cards have more memory, more information, and a higher level of security. They can contain multiple applications on a single chip, manage several passwords, and use authentication and encryption techniques to combine freedom of use with security.

Around the world, there are more than 100 smart card programs in operation. The most common use of smart cards is telephone cards. A major card manufacturer estimates that 420 million phone cards were sold worldwide in 1996. Many leading banks also issue smart cards that function as credit and debit cards. In the United States, as part of its Smart Ship Project, the Department of the Navy uses stored value cards to replace all cash transactions aboard the Aegis class cruiser USS Yorktown. In Mexico, Bancomer, S.A., issues stored-value smart cards to truck drivers for automated toll collection in remote parts of the country, eliminating most error, theft, and abuse in reporting expenditures back to the company. In Thailand, Thai Farmer's Bank issues smart cards that combine a customer's emergency medical records with smart bank IDs and electronic purse features. Eventually, smart card technology may replace cash, checks, and credit cards in transactions that require safety, application logic, and convenience.

Smart cards offer a variety of benefits to consumers. As a replacement for cash, smart cards increase security and simplicity and reduce paperwork and record keeping efforts. You no longer need to carry as much cash; you can print a transaction history at a convenient ATM; and you can avoid time-consuming forms. As a replacement for coins, smart cards increase transaction speed and convenience and reduce bulk. You no longer need to load multiple coins into a vending machine or carry around coins for routine transactions that require exact change.

Smart Card Integrators Inc. (www.s-ci.com) integrated smart card software and hardware with major casino slot machine manufacturers to create a cashless casino system.

Smart cards also offer a variety of benefits to merchants, financial institutions, and other card issuers. As a merchant, you will appreciate higher sales, faster transactions, easier bookkeeping, reduced costs, and fewer losses. Smart card transactions are typically larger and faster than comparable cash transactions. In addition, electronic payments are easier and less expensive to handle and to reconcile than cash payments. As a card issuer, you gain more customers, lower overhead, and decrease fraud through smart cards.

American Express, with its new smart card "Blue," recently has created waves in using smart cards as effective, commonplace, and trusted payment method online. The card, which has an embedded chip, also works as a credit card. But when used in conjunction with sister American Express products Online Wallet and Smart Card Reader, it provides an added two-factor level of security to online shopping enabled by the embedded chip.

The Online Wallet is a free service that stores personal ordering information, such as a Blue card account number, as well as shipping addresses and can automatically fill out online ordering forms. The Smart Card reader attaches to a customer's PC. To make a purchase, a customer inserts the Blue smart card into the reader and inputs his or her personal identification number, or PIN. This process unlocks the Online Wallet for secure online payment. Because customers must have the combination of chip-embedded smart card and PIN to complete a transaction, the Blue card presents an extraordinary level of protection against the types of Internet credit card fraud that concern many customers today. Customers can instantly register for any of these services online at home4.americanexpress.com/blue/card_home.asp.

Today, Blue is the only product to offer a smart card for shopping on the Internet with an additional level of PIN security. When PC manufacturers start embedding smart card readers into computer hardware such as keyboards, more and more credit card issuers are likely to begin offering smart card-enabled security for online purchases.

VeriFone (www.verifone.com) a division of Hewlett-Packard, offers a range of point-of-sale (POS) payment terminals and peripherals, integrated payment software solutions, and Internet commerce solutions. The company continues to explore new payment technologies, refine its product range, and actively promote industry initiatives, including U.S. smart cards and worldwide Internet commerce. The VeriFone Web site includes more information about smart cards and their benefits and how to establish smart card systems.

The Smart Card Industry Association (www.scia.com) is a global trade association whose Web site provides information on smart card technology, technology applications, industry news, and links to other smart card sites, products, and information.

Electronic Billing

Electronic billing lends itself well to both business-to-consumer and business-to-business efforts. Whether you are currently sending out paper-based bills or have not yet developed a billing system, consider the role that electronic billing might play for your organization and its e-commerce solution.

One of the reasons for electronic billing's growing popularity is its clear demonstration of revenue generation. When electronic billing first started gaining momentum, billers and banks were looking at the technology as a way to reduce printing and mailing costs and to speed payment collection. Today electronic billing allows one-to-one marketing and dynamic information presentation. At its full potential, this method of billing allows companies to increase their contacts, deliver customized information, and provide links to other online services.

Electronic billing enables the presentment, payment, and posting of bills on the Internet. Presentment involves taking the static statement data, which is typically directed to printers, and hosting that information on an interactive Web-based bill presentment server. With the Web, merchants can customize the user interface to each individual customer. By processing payments online, electronic billing provides cost savings and cash management benefits that can consummate the collection process directly on the biller's own Web site. After the statement is presented and payment is secured, billing merchants must post to their accounts receivable system and update the account.

A typical electronic billing transaction comprises of six steps:

1. After verifying user name and password, the bill is presented to the customer.

2. The customer reviews, analyzes, and pays the bill. The resulting transaction typically debits either a bank account or a credit card of the customer.

3. The transaction is posted to the clearing house (ACH) flat file.

4. A success or failure response of the verification and posting of the transaction information is sent back to the biller.

5. An electronic customer receipt is generated, indicating success or failure of the transaction.

6. The merchant's clearing house (ACH) file is reformatted and submitted for account settlement.

This burgeoning practice has attracted a variety of companies, each of which is a good source for more information. The large players include Microsoft and CheckFree (www.checkfree.com), a company specializing in payment processing. Other electronic billing vendors include BlueGill Technologies, @Work, Novazen, and eDocs. Service bureaus and billing-service providers, including Pitney Bowes, IBS, and Bell & Howell, ease the transition to electronic billing.

Credit Cards: Web Leaders in Payment Systems

While micropayments, smart cards, and online billing each play an interesting and important role in electronic commerce, the top three credit card companies handle over 98 percent of all purchases made on the Web. Contrast this figure with transactions outside the Internet, where only about 20 percent of all purchases are made with credit cards, and it's obvious that credit cards are the driving force behind online business-to-consumer transactions. Most financial institutions, credit card verification systems, and e-commerce software programs integrate the credit card systems of companies, including American Express, Carte Blanche, Diners Club, JCB, MasterCard, Novus/Discover, and Visa. In fact, MasterCard and Visa account for just over 70 percent of all purchases made by credit cards.

A credit card transaction online is financially akin to any physical credit card transaction. Each transaction is actually a loan made by a lending institution to a customer. A major reason for credit cards' popularity online are consumer protection laws issued by the Federal Reserve

Board, which limit a customer's liability for the fraudulent use of his or her card to $50. Banks sometimes waive this fee if the customer reports a lost or stolen card to them immediately, reducing the potential loss to the bank by allowing quick deactivation of the account. Credit card transactions are handled in much the same way, whether in the physical world or online. The components of a typical transaction include authorization and settlement. Authorization refers to the approval process of a credit card transaction. Settlement refers to the payment of various fees to companies that assist the transaction, as well as payment of the merchant by the bank.

Credit card authorizations can be separated into a series of finite steps. First, a consumer visits a Web site by using a standard Web browser. The consumer selects an item to purchase at the merchant's Web site and adds it to a shopping cart. After entering shipping and credit card information, the consumer is presented with a summary of the item, its price, and the billing information. The payment information is protected by Secure Socket Layer (SSL) encryption and forwarded with the order form to the merchant's commerce site. The server software adds the merchant's identification information to the packet. The secure payment request is forwarded over the Internet and is received through a secure firewall by a series of servers that pass it to the merchant's financial institution. From there, the packet travels to the customer's credit card bank to approve or decline payment authorization. The consumer's credit card bank sends its response back through the merchant's financial institution to the merchant's system. Typically, this authorization process takes 20 seconds or less to accomplish.

After the transaction is authorized, the consumer's credit card bank settles the transaction (a process also known as interchange). As a merchant, you should understand the process by which credit is settled in order to determine if accepting credit cards at your online store makes financial sense and, if so, how you can structure your prices to account for these costs. The merchant delivers the goods to the consumer and requests financial settlement. For a $100 transaction, the merchant receives $98 based on a 2 percent discount rate. That money is deposited in the merchant account at the merchant bank (also known as the acquirer bank). The acquirer bank receives $.67 of the $2 paid by the merchant (the cost of offering credit card service to consumers). The merchant bank pays the association (Visa or Mastercard, for example)

$.08. The issuer of the credit card (issuer bank) receives $1.25 for owning the cardholder relationship. Eventually, the consumer is billed for the purchase and either makes payment or incurs credit card debt. At this point, the entire transaction is complete.

Enable Electronic Payments

Enabling electronic payments over the Internet for your e-commerce business requires establishing an account with a financial institution and an Internet payment company and also setting up your merchant register to accept payments. These payments originate from your customer, who makes purchases by using an electronic wallet or by completing a form with purchase information incorporated into the merchant Web site. The following sections describe each of the components that must be enabled for electronic commerce.

Electronic Wallet

The software component that a consumer uses to make purchases electronically is referred to as an electronic wallet. The wallet allows the customer to store information on his or her desktop, eliminating the need to continually refresh information such as customer name, shipping and billing address, and payment information. Electronic wallets support several types of credit cards, digital cash, paper and digital checks, and purchase orders. Software versions of electronic wallets include vWallet (VeriFone), CommercePOINT Wallet (IBM), and Microsoft Wallet (Microsoft). Most sites do not rely on wallets, because their implementation forces the user to download software and to establish a relatively complicated configuration.

A popular alternative to the electronic wallet is a form that the merchant incorporates into the Web site. This form, filled out by the customer, captures contact information, payment information, and product choices. Most popular commerce solutions, usually via templates, allow these forms to be designed, developed, and deployed easily by the merchant. Although the resulting electronic commerce is form-based, instead of wallet-based, the remainder of the components perform the same functions.

Merchant Register

A merchant register represents the software that businesses use to exchange payment and order information with cardholders and to collect payments from cardholders' financial institution. Features of these registers include support for a variety of payment types, security protocols (SET and SSL), complete integration with a merchant's storefront technology, and financial institution support. A typical merchant register allows the merchant to accept payment via credit cards, micropayments, and smart cards for both time-of-purchase and electronic billing. Most commerce solution vendors allow easy integration of CyberCash and VeriFone products into the register, eliminating the need for extensive programming. In addition, most merchant register software is supported by a wide variety of financial institutions. Versions of merchant registers include vPOS (VeriFone), CommercePOINT eTill (IBM), and CashRegister (Cybercash). Information regarding the products can be found at each vendor's Web site.

Internet Payment Service

An Internet payment service functions as an intermediary between merchant register software and the financial institution. Capabilities, provided by Internet payment services such as CyberCash or VeriFone, enable a merchant to accept online payments from customers. These payments are securely processed from a merchant's storefront through the existing system of financial institutions and credit card processing companies.

When considering various Internet payment services, focus on the key qualities of security, ease-of-use, and price performance. Ensure that the service you select provides security for credit card authentication, transaction confidentiality, server host integrity, and server virus protection. At a minimum, the service should provide Secure Socket Layer (SSL) protocol security and Secure Electronic Transaction (SET) encryption and authentication. An Internet payment service creates an easy-to-use experience by featuring real-time payments, scalability, easy installation and upgrades, and multiple payment methods. Real-time payments mean immediate bank approval of a sale as well as expedient deposit and settlement of payments. Effective scalability allows smooth growth of transaction volume and additional payment features. Determine the complexity and workload required to install and upgrade software and hardware products. Your Internet payment service should allow your customers to change the methods that they use to pay for your products or services. Finally, the company should offer compelling price performance driven by its integration with your storefront and its relationship with a wide variety of financial institutions.

Financial Institutions

A financial institution contracts with a merchant to enable the acceptance, processing, and deposit of credit card transactions. Merchants must maintain an account with a financial institution to receive credit for credit card sales. The financial institution deposits daily credit card sales into the merchant's account, minus any applicable fees, and delivers regular reports based on any transactions.

Acquiring financial institutions sometimes perform merchant services in-house, while others outsource these functions to a third-party processor (i.e., First Data Corporation, Global Payment Systems, Nova Information Systems, or Vital Processing Services). A third-party processor is a company that provides credit card services to merchants on behalf of acquiring financial institutions. Services may include billing, customer support, reporting, authorization, and settlement services for merchants.

Establishing an account with a financial institution is reasonably uncomplicated. Merely contact the institution, describe your e-commerce configuration (what components you are using), and select an Internet payment service. Initially, make sure that the financial institution is authorized to underwrite credit card services to merchants. Direct the financial institution to set up the account to accept credit cards over the Internet using your choice of payment services. Determine the details of establishing a merchant account, including the application and approval process, the fees and charges, and the requirements for opening a deposit account for conducting business. Assume that the entire process will take two to fourteen working days to complete.

Implementing Credit Card Verification

The final phase of enabling electronic transactions is to set up a credit card verification system. Each solution vendor has specific instructions for incorporating verification services into its commerce software. As an example, let us walk through the CyberCash (www.cybercash.com) procedure for implementing credit card verification. CyberCash has just released a service called CashRegister 3, a new Internet payment architecture that makes it easy to integrate storefronts, operate payment services, and upgrade to new services, standards, and options as they become available.

Implementing CyberCash technology begins in one of three different ways. You can use the services of a Merchant Development Partner to integrate and/or to host the CyberCash commerce-enabled Web site for you. You could implement the solution yourself by downloading the CyberCash software and integrating it into your storefront. Or, rather than building a commerce-enabled Web site from scratch, select an e-commerce application or storefront solution in which CyberCash service is already integrated.

In order to accomplish the second option, perform the following steps. Collect and review information about CyberCash and Internet payments (www.cybercash.com). Register with CyberCash, and download the software. Apply for a merchant bank account. The bank accepts your application and then notifies CyberCash. CyberCash enables cash register functionality and e-mails you a notification. Configure your storefront for CyberCash. Then retrieve setup information from Cyber-Cash, and enter it into the storefront. Submit a test transaction. Assuming that everything works, ask to begin service. CyberCash notifies you that service can begin. Now you can open your store for business.

A variety of companies provide credit card transaction solutions for merchants. Bank of America's Merchant Services division is involved in several Internet pilots to facilitate secure payments over the Internet. Charge Solutions provides Internet-based real-time credit card processing to merchants. Credit Card Network, a credit card authorization service, handles credit card transactions for merchants who have already set up their electronic storefront. Epoch Transaction Services offers secure, real-time billing solutions for membership and product-based Web sites. First Data, a huge provider of merchant processing services, handles credit card transactions in the retail, hospitality, supermarket, direct marketing, and health care industries. iBill provides credit card processing services. Paymentech offers full-service electronic payment solutions for merchants, third-party transaction processing, and total commercial card payment programs. Secure Trading, a British company, also offers credit card processing services for Internet electronic commerce sites.

Ensure Online Transaction Security

One form of security, the Secure Electronic Transaction (SET) payment protocol, is an Internet standard designed to provide a high level of security and antifraud assurances for payment card transactions over the Internet. Visa and MasterCard, along with technology partners such as VeriFone, developed SET with the goal of replicating the level of confidence that all parties to a transaction receive when an in-store transaction is conducted. The authentication process involves a series of security checks performed using digital certificates issued to customers, merchants, banks, and payment brands.

The four SET components include a cardholder wallet, a merchant server, a payment gateway, and a certificate authority. The cardholder wallet generates SET protocol messages that can be accepted by the three other components. The merchant server component processes payment card transactions and authorizations. The payment gateway component (run by an acquirer or a designated third party) processes merchant authorization and payment messages and interfaces with private financial networks. The certificate authority component issues and verifies digital certificates as requested by the other components.

A transaction secured by a SET-enabled browser and a SET-enabled server follows these steps. The customer opens a bank account that features Mastercard or Visa services. The customer receives an electronic file (a digital certificate) that functions as a credit card for online purchases. Third-party merchants also receive digital certificates from the bank. The customer then places an order, either over the Web, via e-mail, or by telephone. The customer's browser receives and verifies the validity of the digital certificate from the merchant. The browser then sends encrypted purchase details. The merchant receives and verifies the validity of the digital certificate from the customer. The merchant then sends to the bank the order information, including the bank's public key, the customer's payment information, and the merchant's certificate. The bank then verifies the merchant certificate and the order information message. The bank digitally signs and sends the proper authorization to the merchant. Finally, the order is filled.

The SET protocol provides confidentiality, data integrity, verification of identity, and nonrepudiation of disputed charges. The SET protocol uses advanced public key cryptography algorithms to assure that messages are transmitted and received confidentially—safe from the prying eyes of any uninvolved parties on the Internet. Cryptography algorithms generate the encryption (scrambling) and decryption (unscrambling) of information. The process of public key cryptography allows anyone to send an encrypted message to a designated recipient, using what is known as a public key. The recipient then uses a private key to decrypt the message. Therefore, only the designated recipient has the ability to read the message.

The SET protocol can also send credit card information directly from the consumer to the merchant's financial institution, without allowing the merchant access to the cardholder's account information. A consumer can therefore make a purchase with confidence that no fraudulent activity will take place with his or her credit card number after the card number has been transmitted. Merchants accepting SET transactions in this manner are protected from the liability of credit card numbers in their database, because they do not receive the credit card information.

When a message sent via the SET protocol is encrypted, a unique numerical value is generated for that message. If anyone attempts to tamper with the data contained in the message, the numerical value for the message would change and would therefore be detected. This process ensures the integrity of the message.

The SET protocol uses digital certificates to verify the identity of all parties involved in a transaction. When a consumer makes a purchase using the SET protocol, a digital certificate is sent from the consumer to the merchant and, simultaneously, from the merchant to the consumer.

The consumer's digital certificate functions as a signed credit card slip to the merchant. The digital certificate provides verification of the consumer's identity and offers a high level of assurance that the credit card is not being used fraudulently by the consumer.

The merchant's digital certificate provides the consumer with verification of the merchant's identity and leaves the consumer with a high level of confidence that the merchant is operating a legitimate business and that the purchased items will be shipped.

The process of digital certificate distribution that takes place during SET transactions also holds the consumer and merchant accountable for information contained in the transaction. Therefore, if a consumer places an order for a product using SET and then receives the product, he or she cannot deny that the order has been placed. Similarly, merchants cannot later deny that they received the order.

Additional Security: The Secure Sockets Layer

Another form of security, Secure Sockets Layer (SSL), is a commonly used program layer for transmitting payment information between a buyer and a seller. SSL, which was developed by Netscape, sends data back and forth between the cardholder's Web browser software (such as Netscape Communicator or Microsoft Explorer) and the merchant's Web server software (such as Netscape's Enterprise Server or Microsoft's Commerce Server), within a secure environment. SSL encapsulates data into transactions and guarantees the confidentiality and integrity of the individual blocks of the transaction.

The "sockets" part of the term refers to the sockets method of moving data back and forth between client and server in a network. A socket is defined as the endpoint in a connection. Passing sockets is a method for communicating between a client program and a server program in a network. Sockets are created and used with a set of programming requests (or function calls), sometimes called the sockets application programming interface (API).

The "layer" part of the term refers to the program layer between an application and the Internet's TCP/IP layer. The programming code for keeping your messages secure is contained in the program layer.

SSL uses the public key/private key encryption system from RSA, which also includes the use of a digital certificate. The secure environment for SSL is created through the use of public key cryptography, which consists of the encryption (scrambling) and decryption (unscrambling) of information. Public key cryptography allows anyone to send an encrypted message to a designated recipient, using what is known as a public key. The recipient then uses a private key to decrypt the message. As a result, only the designated recipient has the ability to read the message.

The SSL protocol is a powerful tool for the secure distribution of information but does not address all of the risks associated with sending and accepting transactions over the Internet. For example, SSL establishes a secure session between a browser and a server. During the period when the browser is logged onto an SSL server, authentication between the browser and the server takes place. However, SSL does not authenticate the parties who are using that software. Thus, while cardholders using SSL can submit payment information free from the prying eyes of a third party, there is no way of verifying the identity of the online storefront that they are visiting.

Help Customers Trust You

Whether you are a buyer or a seller, the initial requirement of any real-world transaction is identification. Answering the question, "Who are you?" is essential. When you, as a shopper, walk into a franchise hardware store, let's say Home Depot, you already are comfortable with the brands offered, the service you receive, and the store's return policy. You do not enter the store wondering whether you could be subject to theft or fraud in the process of your shopping experience there. Instead, it is more likely that Home Depot may wonder who you are and whether you may subject the store to theft or fraud.

The tables of suspicion turn when shoppers go online. Once online, the shopper questions the merchant's integrity, and the merchant feels confident because it has established layers of security to protect its interests prior to opening for business. Therefore, from a merchant's perspective, establishing the trust of customers is critical to the success of any e-commerce solution.

> *Encouraging trust in e-commerce*
> *involves six primary components:*
> *seals of approval, brands,*
> *navigation, fulfillment,*
> *presentation, and technology.*

Encouraging trust in e-commerce involves six primary components: seals of approval, brands, navigation, fulfillment, presentation, and technology.

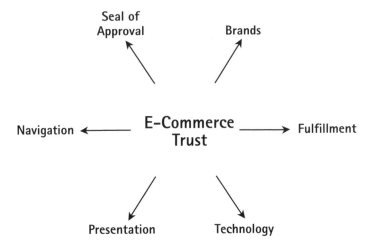

Figure 7.5 The primary factors for gaining trust from customers.

Seals of approval, represented by symbols like VeriSign, are designed to reassure visitors that sites have established effective security measures. Known as security brands, these seals of approval testify to the safety of a merchant's site, its technology, and the network behind it. Consumers recognize these signs and feel more comfortable in a merchant's identity knowing that merchant is affiliated with these known vendors.

Brand refers to a company's implicit promise to deliver specific attributes, based on a company's reputation and a visitor's previous experience with its products. The concept of branding, which we covered in Chapter 3, includes online and offline brand recognition, portal or marketing affiliations, community building, and the sense of a site's breadth of product offerings.

Navigation refers to the ease of finding what a visitor is seeking. Navigation is aided by understandable terms, the consistent placement of navigation elements, clear instructions to help shoppers make their way through a site, and simple, appropriate terms to describe site content.

How clearly a site indicates the way orders are to be processed, its return policy, and how well it explains the way customers can seek recourse to problems are all elements associated with fulfillment. The assurance that a customer's personal information will be kept secure and private is a key attribute of fulfillment.

Presentation implies the ways in which the look of a site communicates meaningful information. On the home page, a site's purpose must be clear to the first-time visitor.

Visitors evaluate technology largely in terms of speed and function. They notice how well a Web site's technology operates and how quickly each page loads. Therefore, it is important to balance page-loading speeds with helpful functionality.

In order to address each of these components and to establish customers' trust, begin by adding a policy statement explaining how transactions work and how returns are handled. Provide your corporate history. Add a variety of security seals, including VeriSign and the Better Business Bureau. Display a message indicating that any transactions that take place are secure and encrypted. Also, provide a toll-free number to allow users to pay by phone if they choose. When designing the security features of your e-commerce site, assume the perspective of customers and address their needs.

8

Know Your Infrastructure

Even with the emphasis placed on the business focus of e-commerce solution development, an understanding of the underlying technologies is critical. Few individuals will actually have responsibility for implementing or installing the software required to deploy a solution. However, knowledge of product functionality, limitations, and technical requirements offers a measurable competitive advantage.

> *Knowledge of product functionality, limitations, and technical requirements offers a measurable competitive advantage.*

Overview of Components

The components that comprise the typical electronic commerce solution are Web clients, Web server software, commerce server software, connectivity tools, and back-end systems. Learn the features and functions of each of these components, as well as how they relate to each other. In later chapters, we address vendor solutions, payment systems, and security considerations, with each chapter focused on the component nature of a solution. The vocabulary associated with these five components enhances your understanding of the design process and proves invaluable as you select software and hardware packages for implementation. Figure 8.1 illustrates a typical e-commerce architecture.

Web server software publishes hypermedia documents on the fly for the Web client interface. Commerce server software generates and supports storefront functionality for merchants. The Web server software and the commerce server software reside on the same hardware component. Customers access the vendor site via Web clients (or, as they are more commonly known, Web browsers). Back-end systems, which typically already exist in an organization, require a connectivity tool to link them to the server software. This linking allows companies to leverage data such as product information, customer information, and transaction records.

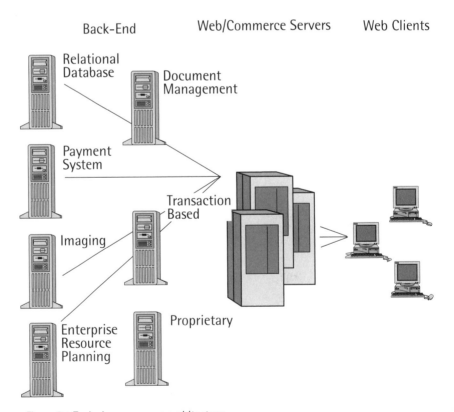

Figure 8.1 Typical e-commerce architecture.

E-commerce is historically a direct descendent of electronic data interchange (EDI). This technology is used to communicate between large companies. However, standards, based on the specific industry, are required for the exchange of information (data) electronically between computer systems. For example, the automotive industry requires that each participant use the Automotive Industry Action Group (AIAG) standard to communicate. Grocery retailers are required to use the Uniform Communications Standard (UCS). Furthermore, all participants are required to be connected to the same private network. So, how do the smaller companies participate? This is one of the major problems. EDI is very costly to implement, and it is difficult to provide interoperability between different industries, because each industry has its own standard for data exchange.

For companies that currently communicate via EDI, e-commerce software vendors provide solutions that allow the integration of existing EDI systems. Microsoft, IBM, Lotus, Netscape, and other software vendors provide EDI extensions as part of their tool sets to integrate into existing EDI-based systems. As part of these solutions, companies that were restricted to automating processes between business partners that follow the EDI standards can now send information, such as purchase orders, directly to smaller business partners via a Web browser.

The electronic commerce architecture, in comparison, allows companies to communicate using one standard communication protocol, called TCP/IP (Transmission Control Protocol/Internet Protocol), and a data format called HTML (HyperText Markup Language). Think of a protocol as a language. In order for two people to speak and then understand each other, they need to communicate in the same language. What happens when one person speaks and understands only German and the other speaks and understands only Italian? They need a translator. The communication between these individuals requires some additional time and effort. In addition, the transaction costs a little more, because the translator must be paid. That is the beauty of communicating in one common language . . . no additional translation cost. Electronic commerce has enabled interoperability between different business systems, increasing efficiency and lowering costs. Businesses can now easily communicate with each other through Web servers and with consumers through Web browsers. That is why we are seeing a big shift in the way business is being conducted on a city-, state-, country-, and worldwide basis. Everyone speaks the same language.

Web Clients

A Web client (from the perspective of the merchant) or browser (from the perspective of the customer) is an application program that provides a graphical interface to view and interact with all the information available on the World Wide Web. The word *browser* originated as a generic term for user interfaces that let you browse text files online. As Web content began to also include graphics, the term became more common.

Technically, a Web browser is a client program that uses the Hyper-Text Transfer Protocol (HTTP) to make requests of Web servers throughout the Internet on behalf of the browser user. A commercial version of the original browser, Mosaic, is still popular today. This version, including many of the user interface features in Mosaic, is Netscape Navigator. As Microsoft began to realize the growing popularity of the Web, it developed and freely distributed Internet Explorer. Although online services such as America Online, CompuServe, and Prodigy, originally featured their own browsers, virtually all providers now offer the Netscape or Microsoft browser. Other common browsers include Lynx, a text-only browser for UNIX shell and VMS users, and Opera, a sophisticated, yet compact, Web browser.

An important component of the development process for an electronic commerce solution is testing your resulting Web site pages, embedded code, and scripts on the dominant browsers available to the public. Currently, those browsers are Microsoft's Internet Explorer and Netscape's Navigator (or the Communicator suite). Microsoft commands a 73 percent market share to Netscape's 25 percent with 2 percent shared among several others. Statistics on browser usage are available and updated daily at www.statmarket.com. Always build solutions for the latest proven version of the de facto standard browser.

Web Server Software

Web server software serves as a liaison between back-end systems and front-end Web clients. Its primary function is to generate and deliver hypermedia documents based on HyperText Markup Language (HTML). In most cases, the HTML code is generated on the fly by the Web server software. The World Wide Web Consortium (www.w3.org)

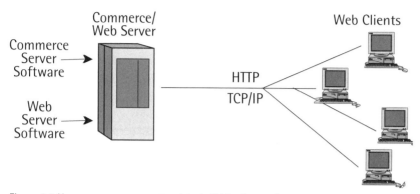

Figure 8.2 How server components relate to Web client software.

sanctions the latest standards of HTML. Vendors of these Web server products support a variety of network operating systems, including NT, UNIX, Novell, and OS/2. Web server candidates for your electronic commerce solution include Microsoft Internet Information Server (IIS), Domino Server, Netscape, and Apache Server. Figure 8.2 illustrates the Web server/commerce server components and their relationship to Web client software. For more information about Web server software and its functionality and suitability, visit the major vendors' Web sites.

Commerce Server Software

At the heart of an e-commerce solution lies the commerce server software, with capabilities for creating an online storefront. Several major vendors on the market offer a variety of software solutions that allow you to establish an e-commerce site on the Web. Most vendors offer business-to-consumer features and functions as part of the template. Other vendors focus on solutions that provide options to set up net marketplaces, business-to-business sites, supply-chain management, and procurement.

The storefront is the focal point for selling your products and services. Storefronts may incorporate and support products for sale, services provided, company information, job postings, special events, featured products, store registration, and other elements that add value to your customers' experiences at your site.

Selling Products

The most common shopping model on the Web involves using storefronts to sell products. When buying products, the customer selects the item to purchase and adds it to a shopping cart. Once the customer has finished shopping, he or she pays for the item online using a credit card. After the credit card is approved, the system sends the customer a confirmation, and the items are delivered. The concept is very similar to the product catalogs currently used by thousands of companies. Each product has a product SKU, name, description, and price. Other features may include color, size, weight, images of the product, or any other attribute that is specific to the product you are selling.

Selling Digital Content

As bandwidth available to the consumer increases, electronic delivery of digital content becomes more practical and more popular. Most commerce software does not provide an "out-of-the-box" solution to support the sale and delivery of electronic goods that customers download. However, the major solutions do support the purchase side of the transaction. Typically, a third-party component is required to support the delivery side of the transaction. This chapter addresses only the purchase side of the transaction, a common ground for the major commerce software components.

When initially creating the storefront, standard wizards are used to select your store model. Several vendors have sample stores that you can use as a basis for the site. The selected template serves as a model for your eventual store and represents the hub of your e-commerce site.

After you walk through the wizard screens, the site still requires modification. The most common question about these wizards is whether people really use them. The answer is yes, if they are creating a storefront to sell products or services. The rationale is that these wizards contain the basic information you need to create the site, such as the storefront home page, product information, tax and shipping information, credit card information, design, and layout. These wizards also allow you to easily integrate third-party credit card verification systems (such as Cybercash or VeriFone), Taxware International services, UPS tracking capabilities, and inventory tracking systems. The integration of these third-party components represents the most compelling reason to at least consider the wizards.

Why reinvent the wheel? It is much easier to create a store using these tools than it is to construct a store from scratch. If you build your solution from the ground floor up, it may take several developers several months to reach the same outcome offered by the wizard-based solution. Put your efforts where your expertise is most developed. Where it is less developed, harness the expertise of others. In the field of e-commerce, time is your primary opponent. Do not spend time learning a skill that you can rent instead.

Why reinvent the wheel?

When selecting a commerce solution, it is important to review the built-in templates or wizards that are part of the product. Each vendor has solutions that vary slightly in their out-of-the-box capabilities. By purchasing a package that meets your business needs, you can reduce the amount of time and resources required to customize a solution. You must compare the cost of a standard software solution with that of a custom solution designed and developed by internal or external resources.

After you have designed, developed, and deployed your store, your focus falls on administration of the store's products. The easier, the better. Several e-commerce solutions allow you to tweak the commerce component of your site via any common Web browser. Administration includes capabilities such as changing product properties with point-and-click expedience, as well as adding, updating, and deleting products. The ability to track inventory varies from solution to solution. Some vendors offer the capability of integrating your commerce site into existing inventory tracking systems. However, most vendor solutions stop short of supporting order tracking. What happens to stock on hand after an order? To take the next logical step, most systems require customization to tie into existing inventory tracking systems. As support for this integration seems to be a weak point with most of this commerce software, you are likely to be forced to build your own solution.

Your solution, because it is developed from the ground floor up, may track incoming products, items ordered by customers, and stock on hand. For new companies, you need to create your traditional and your online transaction systems. This means creating additional tables to store data, integrating the items ordered from the online site, building an inventory system to track items ordered, assessing stock on hand, determining low supply of stock, and establishing reorder levels.

Purchase order (PO) generation is a standard capability for most commerce packages, but some require more customization than others. This feature is important in business-to-business commerce where most companies work with PO instead of credit card transactions. One of the weak links of the typical PO system is automating the updating process of internal systems between companies.

Although the majority of payments at Web-based stores are made through traditional credit card transactions, you need to be in position to accept as many payment methods as possible in order to maximize your sales. The most common credit cards accepted by these packages are MasterCard, Visa, American Express, and Novus/Discover. However, you can also customize a package to include other credit cards. On the credit verification system side, Cybercash and VeriFone represent the two vendors that usually provide easiest integration into commerce software. Of the two systems, Cybercash dominates the market. Other payment methods include micropayments, smart cards, and online billing systems. Choose a solution that supports a wide range of electronic and traditional payment methods. For more information, see the Payment Systems chapter.

Most commerce packages include the option to specify sales tax and shipping costs. Calculating a customer's sales tax and shipping costs is a vital component of any good e-commerce software package. Customers want to know their total cost prior to placing their order. Shipping costs need to be calculated in real time. Further complicating this issue, a significant proportion of orders in your Web store may come from international customers. Select a solution that allows you to calculate and specify shipping costs with a great amount of detail.

Remember that taxes vary by state and by nation. How do you address this element of your commerce package? Third-party tax and accounting packages such as Taxware International or Great Plains Accounting Software allow you to integrate their product into your Web site. Even some of the store design wizards support this integration.

Site analysis provides you with the option to track visitors to your site, products purchased, and pages accessed. Because several of these packages have thousands of items you can analyze, we recommend spending some time planning prior to setting up the tools. Decide exactly what you want to monitor. If you do not plan out exactly what type of trends and analysis you are trying to capture, the analysis tools available could overwhelm your efforts. The goal is to produce valuable information, not just data. And for all you technologists reading the book, ask the marketing, sales, and product development departments what information is valuable to them for their work.

Commerce packages vary on how much they offer for site analysis. The packages that do not provide the features you need typically have third-party solutions that you can substitute. It is important that you review site analysis results to determine which advertising campaigns are generating sales, which pages are not accessed, and, finally, how effectively your site supports your e-commerce ambitions. You can use this information to customize your site for an optimal return on your Web site investment.

Order Tracking

Another standard feature of commerce software is order tracking, both internally for your company and externally for the consumer. Order tracking allows you to view pending, disapproved, and processed (completed) orders. Typically, once a customer's credit card is approved, the status of an order changes automatically from pending to approved. For purchase orders, the process may include a combination of automated approvals and manual approvals. The goal, however, is total automation of the process. Once an order is approved, the order is completed, and the product is shipped to the customer.

Most commerce packages provide an option to allow customers to track their packages via UPS or Federal Express (FedEx). For example, tracking is part of UPS Internet Tools, which is designed to help merchants serve their customers better. UPS Internet Tools allows customers to track their packages worldwide without leaving your commerce site. UPS tracking is available in local languages in sixteen countries and territories via the Internet. Tracking is also available via the UPS home page.

FedEx, the other dominant shipper, is now offering a set of tools to provide shipping and tracking functionality. FedEx ShipAPI provides you with a shipping application programming interface (API), sample templates, and instructions on how to connect seamlessly to FedEx. It streamlines your online shipping process by integrating FedEx shipping templates and tools into your company's site or corporate information system. You can connect directly to FedEx when placing shipping orders and scheduling pickup requests.

You can also let customers check on the progress of a shipment instantly from their own Internet-enabled computer. You can track multiple packages simultaneously, consolidating several inquiries into one simple transaction. You can access available shipment information, using ship date and country code or account number to check progress. You can customize sample templates for tighter integration into your online environment. And, finally, you can retrieve and store additional shipment information to provide your users instant access to detailed historical shipping data. Built-in error-checking capabilities verify the accuracy of required tracking information. The tracking system uses public-key encryption software to help protect confidential data information, as well as to verify data integrity and authentication.

As the popularity of the World Wide Web gains momentum, the amount of information available online rises dramatically. Site visitors experience increasing difficulty in finding specific information. If site visitors cannot locate the information they want (and quickly), they move on to another site. Commerce solutions provide site administrators with tools to gather and index information and to create key word search capabilities on your site.

Keyword search capabilities enable site visitors to use highly targeted search queries to find and retrieve the specific information they need. You can gather and index information by crawling the Internet, an intranet, Exchange Public folders, NNTP newsgroups, an ODBC database, or a file system. Document properties are the characteristics of the document, including the author's name, date created, title, subject, documents address, and contents of the document. Once extracted, the document properties are stored in a central location.

In addition, search tools provide your site visitors with a Web page from which they can initiate a search and a page to display the results of the search. These pages can be extensively tailored to the needs of your site visitors and to the information that is available in your catalogs.

Customer Registration

Customer registration is also a standard feature for commerce software solutions. Registration may be set up as soon as visitors access the site, as soon as visitors want to purchase a product or service, or when visitors access confidential information. For the first scenario, where access to information is dependent on the user login name and password, multiple vendors can access your site and be provided different products or services to purchase. An example of the second scenario is *The Wall Street Journal,* which requires that you register and pay a fee when accessing the newspaper online. An example of the third scenario is American Express, which offers online options to view credit card information. The site requires that visitors enter information prior to accessing their personal accounts.

Another reason that registering visitors is extremely important to your company is that it allows you to capture data about your customer. This data can later be used to analyze buying trends, send product information back to your customer, and set up personalization of options based on their preferences.

Cross-selling features use shopper trends and existing transaction information to recommend products of interest to customers. For example, Amazon recommends additional books to customers based on their previous purchases or based on a book they are browsing.

Think about visiting a store. The natural links between products arise almost automatically. When you visit a camera store, the store owner strategically places all of the other associated items that you may need: film, batteries, flash, photo albums, and the like. By the time you finally leave the store, you have spent twice what you originally intended. As a store owner, you should offer products associated with the products that your customer seeks. You provide convenience. The customer provides additional revenue.

Once your store is operational, you need the option to run promotions. You should be able to quickly change a price or offer discounts for volume purchases or to special customers. Advertising capabilities may include banners, featured products, and special discounts. Some of these packages even provide options to track how many clicks have been made on each banner and allow you to set up the banners and ads so that when a visitor searches for a specific piece of information a related banner is displayed. Other types of ads included are "buy now" ads for impulse buyers. Some commerce software includes great options for advertising. Other commerce software offers very little, requiring that you purchase a third-party advertising plug-in or build the functionality into the site.

Push technology provides users with channels that deliver specific content. The user subscribes to a channel and related information is pushed (delivered) to the browser on the desktop. Most people have heard of PointCast, one of the first push technologies out on the market. Now commerce vendors provide solutions featuring push technology as one of their delivery mechanisms.

Direct mail is similar to an electronic version of paper-based direct mail. The difference is that you can use direct mail to deliver personalized content to thousands of users with lower costs than paper-based mail. In addition, direct mail offers automated delivery (a standard of marketing).

Commerce server management tools may be included as part of the commerce software or as part of the management software provided with a commerce solution. Most vendors include the following capabilities: content management, replication and clustering, site usage statistics, and remote administration.

Content Management

Content management is a feature that allows you to manage your content, including new products, the design of pages, and additional categories. Think of content as all of the information displayed on the site prior to that information being displayed to the world. Content management is critical to the success of any site. The world is full of visually stunning sites. However, sites often feature misspelled words, poor grammar, and incorrect data. Remember that your customer's first impression could become a last(ing) impression.

When selecting a commerce package, ensure its content management capabilities. This includes the ability to review your information prior to replicating it to the servers that your customers or suppliers access. Typically, content management involves a central hub server where all changes are made. Approvals are set up by assigning privileges to specific employees within the company. These employees are required to approve the pages prior to sending them out to the external servers. If they are not on the list as approvers of the information, they do not have the right to modify the information. Once approved, the pages are replicated out to the production servers. This review should include both the Web page design and the specific contents.

Figure 8.3 illustrates a typical content management process. The author of a document sends (1) the document to the hub server, the person responsible for authorizing the document reviews (2) the docu-

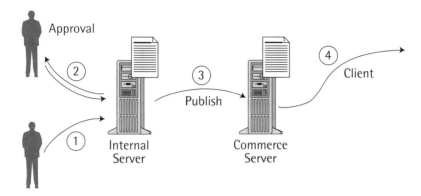

Figure 8.3 A typical content management process.

ment needed to be approved. Once approved, the document is sent (3) to the external commerce servers. Clients retrieve (4) the document from the commerce server.

Connectivity Tools

Connectivity tools act as translators to connect back-end systems to server software or front-end clients (browsers). Early e-commerce Web sites required the use of programming languages, such as C++, to perform the appropriate data translations that link these components. Today, many commerce software providers and third-party vendors provide tools that offer connectivity with minimal programming efforts. These tools are designed typically to push and pull data from the Web/Commerce server to the back-end system.

Back-End Systems

When designing and deploying an e-commerce solution, make sure that you can easily integrate your existing back-end systems.

Make sure that you can easily integrate your existing back-end systems.

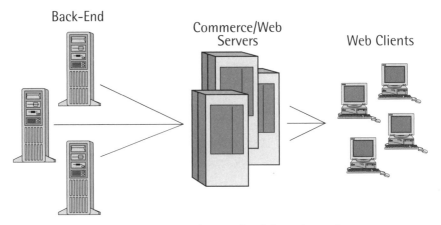

Figure 8.4 Connectivity tools are needed to translate information to the
front-end client.

Determine exactly what systems you have in place. Your commerce
solution requires that you pull data from these systems to your commerce site. Gather the names of the software vendor and the software
packages (with version numbers). In addition, you need to understand
your current infrastructure. Back-end systems may include relational
databases, transaction-based systems, ERP systems, EDI, third-party
software, and proprietary systems.

Integration with existing relational databases is provided in most
commerce package solutions. Each vendor has its own set of tools that
provides integration to specific relational databases. The question you
must answer is how easily a commerce system can be integrated. Commonly supported databases include Microsoft's SQL, IBM's DB2, and
Oracle's relational databases. Even if there is not an out-of-the-box solution, some other database vendors, such as Sybase, have built tool sets
that you can use to tie into these systems. Some database vendors have
code at their site that you can use as a starting point to tie into these systems. Begin your research at the commerce system vendor's site to determine if they include a tool to support integration.

The goals of the integration of commerce systems with transaction-
based systems are seamless processing of orders and timely, accurate
updates. Most commerce vendors offer connections to a variety of
transaction-based systems, although with varying degrees of compati-
bility. The questions you must address in your research include how difficult the implementation is and how robust the solution is.

The focuses of IBM and Lotus include integration into any back-end environment, especially legacy platforms (CICS and IMS) and transactional middleware (MQSeries). Their solutions have grown more robust and user friendly with each release. Integration tools include database tables, scripts, sample code, and documentation to support transactions and message queues with other environments. In contrast, some vendors, including Microsoft, have not focused on providing the same breadth of connectivity.

Most vendors provide solutions to integrate with some of the most common enterprise resource planning (ERP) systems, including People Soft and SAP. Commerce server systems add value to your existing ERP system by passing only fully formatted orders, by acting as a firewall, by eliminating direct access of Internet users, and by enhancing the catalogs presented to customers with more interactivity.

If you have an accounting package, tax software, or any other type of system that you want to integrate into the commerce environment, go to the commerce vendors and the third-party vendors to determine if an integration tool is available. If it is not, you must write the API to perform the integration.

Electronic data interchange (EDI) allows program-to-program data exchange over private or public networks. It is based on a set of predefined relationships and standards between the participating parties. These predefined relationships and standards include interfaces, translators, consistent data, and mapping protocols. Commerce server solutions support electronic data interchange by delivering merchant orders or customer special orders to the supplier and by allowing suppliers to report order status to the merchant.

Connecting a commerce system to a proprietary software system requires building an application programming interface (API). An API serves as a liaison by making requests between the two systems. For example, many companies use a proprietary credit card verification system. An API connects the commerce server to that proprietary system, ensuring efficient and accurate communication.

Tuning and Load Balancing

Your Web site might perform adequately as soon as it is deployed. It might continue to function effectively for days or even weeks. At some point, the demands made on the system will overtax it, reducing its performance. Servers deliver Web pages with increasing delay. Some pages are not even available. Site traffic bogs down the system. And then the traffic vanishes. Customers, frustrated with the delays and navigational obstacles, stop visiting your site.

The performance of your electronic commerce solution is typically measured by its fault tolerance and load balance. Fault tolerance refers to the capacity of a site to continue to provide service, even when a server node fails. A fault tolerant configuration features server nodes that immediately pick up the requested load with minimal disruption to visitors when a server node stops working. The process of automatically transferring workload from one server node to another server node is called failover. The process of restoring the workload back to the original server node is called failback. Failover and failback are processes that clustering applications perform automatically and transparently, so that the visitor is typically unaware of their occurrence.

Load balance indicates that the amount of traffic on an individual server within a group of servers is relatively equal to the demand placed on each of the group members. Rather than overloading one server node and underloading another server node, the load can be handled efficiently by distributing the load between the two. Load balance can be achieved by assigning Web sites to a specific server node manually. This configuration allows a Web site to be accessed from either node, although a specific Web site may not be accessed from both nodes simultaneously.

Clustering supports load balance of a system, as well as fault tolerance capabilities. Replication supports quicker retrieval of data by users and fault tolerant operation of the e-commerce solution. Software vendors incorporate both of these features into most commerce packages. An additional component (either software or hardware), called a router, helps balance the load demand for individual servers. The following section describes each of these enhancements to your system.

Clustering

Clustering allows two servers, with real-time updating, to appear as one to site visitors. The servers are connected not only physically, by cables, but also programmatically, with clustering software. This connection allows the servers to support failover and failback functionality

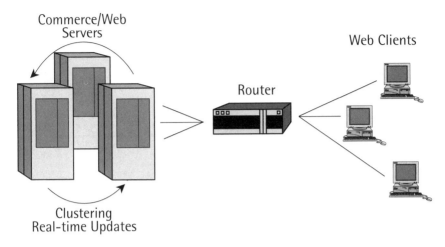

Figure 8.5 Clustering and routing software support load balance of a system.

as well as load balancing. Clustering software is typically included with the commerce package.

Replication

Replication refers to the copying of configuration information and content from one server to another so that both servers offer the same resources to visitors. The site administrator determines and schedules the frequency of further updates of data. The advantages of replication are improved performance and better availability. Applications can operate on local copies instead of having to communicate with remote sites, improving performance. A replicated server remains available for processing as long as at least one copy remains available. The primary disadvantage is that when a replicated server is updated, all copies must be updated. Replication is supported by most commerce packages.

Routers

In both of the previous diagrams, a router sits between Web/Commerce servers and Web clients. Although these diagrams are simplified (packets of information travel to and from other servers across

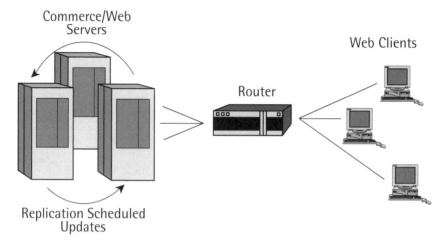

Figure 8.6 Replication improves server performance.

the Internet before reaching clients), the position of the router is accurate. A router is a hardware device or, in some cases, software incorporated into the server, that determines the next network point to which a packet of data should be forwarded toward its destination. The router is connected to at least two network pathways and determines which pathway to route each information packet based on its understanding of the state of the networks.

A router creates or maintains a table of available routes and their conditions. The router then combines this information with distance and cost algorithms to determine the best route for a specific packet. From the perspective of the commerce package, the advantage of routers is their load balancing functionality. Leading manufacturers of routers include Ascend Communications, Cisco, and 3Com. More information about routers and their implementation is available at the vendors' sites.

Network Architecture

While the concept of architecture within this chapter primarily refers to the functionality and relationship of electronic commerce components, a higher-level perspective concerns the actual configuration of a network. Any network can be classified as one of three arrangements: Internet, intranet, or extranet. The significance of this classification is that it establishes the audience, security considerations, and performance levels that your commerce solution must address.

Internet

The Internet, usually called the "Net," is a worldwide system of computer networks, a network of networks in which users at any one computer can, with permission, get information from any other computer. It was conceived by the Advanced Research Projects Agency (ARPA) of the U.S. government in 1969 and was first known as the ARPANet. The original aim was to create a network that would allow users of a research computer at one university to be able to communicate with research computers at other universities. A side benefit of ARPANet's design was that, because messages could be routed or rerouted in more than one direction, the network could continue to function even if parts of it were destroyed.

Today, the Internet is a public, cooperative, and self-sustaining facility accessible to hundreds of millions of people worldwide. Typical users include company employees, government workers, researchers at educational institutions, and private individuals. Physically, the Internet uses a portion of the total resources of the currently existing public telecommunication networks. Technically, what distinguishes the Internet is its use of a set of protocols called TCP/IP (Transmission Control Protocol/Internet Protocol). Two recent adaptations of Internet technology, the intranet and the extranet, also use the TCP/IP protocol.

The most widely used part of the Internet is the World Wide Web (often abbreviated "WWW" or called "the Web"). Its compelling feature is hypertext, a method of instant cross-referencing. In most Web sites, certain words or phrases appear in text of a different color than the rest and underlined. When users select one of these words or phrases, they are transferred to the site or page that is associated with this word or phrase. Sometimes there are buttons, images, or portions of images that are "clickable." If you move the pointer over a spot on a Web site, the pointer changes into a hand, indicating that you can click and be transferred to another site.

The Web allows access to millions of pages of information. Web surfing (moving from page to linked page) requires a Web browser, the most popular of which are Netscape Navigator and Microsoft Internet Explorer. The appearance of a particular Web site may vary slightly, depending on the browser used. Also, later versions of a particular browser are able to render more bells and whistles, such as animation, virtual reality, sound, and music files, than earlier versions.

Intranet

An intranet is a series of networks contained within an enterprise. It may consist of many interlinked local area networks (LAN) and also relies on leased lines in the wide area network (WAN). Typically, an intranet includes connections through one or more gateway computers to the Internet. The main purpose of an intranet is to share company information and computing resources among employees. An intranet can also facilitate working in groups and teleconferences. Web browsers allow companies to deploy the intranet infrastructure by providing a graphical user interface (GUI) that is inexpensive and easy to deploy.

An intranet uses TCP/IP, HTTP, and other Internet protocols and in general resembles a private version of the Internet. With tunneling, companies can send private messages through the public network with special encryption/decryption and other security safeguards to connect one part of their intranet to another part.

Typically, larger enterprises allow users within their intranet to access the public Internet through firewall servers that have the ability to screen messages in both directions to maintain company security. When part of an intranet is made accessible to customers, partners, suppliers, or others outside the company, that part is an extranet.

Extranet

An extranet is a private network that uses Internet protocols and the public telecommunication system to securely share part of a business' information or operations with suppliers, vendors, partners, customers, or other businesses. Think of an extranet as part of a company's intranet that is extended to users outside the company. The same benefits that HTML, HTTP, SMTP, and other Internet technologies have brought to the Internet and to corporate intranets now facilitate business-to-business transactions.

An extranet requires security and privacy. These characteristics require firewall server management, the issuance and use of digital certificates or similar means of user authentication, encryption of messages, and the use of virtual private networks (VPNs) that tunnel through the public network.

Companies can use an extranet to exchange data using electronic data interchange (EDI), allow access to product catalogs exclusively for wholesalers, collaborate with other companies on joint development efforts, and develop and use training programs with other companies. They can also provide or access services provided by one company to a group of other companies, such as an online banking application managed by one company on behalf of affiliated banks, and share news of common interest exclusively with partner companies.

Architecture Security

Protecting your system components typically requires both firewalls and proxy servers. Firewalls protect the resources of a private network from users of other networks. Proxy servers ensure security, administrative control, and caching service. A proxy server is associated with or part of a gateway server that separates the enterprise network from the outside network and a firewall server that protects the enterprise network from outside intrusion.

FIREWALLS   A firewall is a set of related programs, located at a network gateway server, that protects the resources of a private network from users of other networks. The term also implies the security policy that is used with the programs. An enterprise with an intranet that allows its workers access to the Internet installs a firewall to prevent outsiders from accessing its own private data resources and for controlling what outside resources its own users can access.

A firewall, in conjunction with a router program, filters all network packets to determine whether to forward them toward their destination. A firewall also includes or works with a proxy server that makes network requests on behalf of workstation users. A firewall is often installed in a specially designated computer separate from the rest of the network so that no incoming request can directly access private network resources.

There are a number of firewall screening methods. A simple one is to screen requests to make sure they come from acceptable, previously identified IP addresses and domain names. For mobile users, firewalls allow remote access into the private network with secure logon procedures and authentication certificates.

A number of companies make firewall products. Features include logging and reporting, automatic alarms at given levels of intrusion, and a graphical user interface for administrators.

PROXY SERVERS　For an enterprise that accesses the Internet, a proxy server is a server that acts as an intermediary between a workstation user and the Internet so that the enterprise can ensure security, administrative control, and caching service. A proxy server separates the enterprise network from the outside network, while a firewall server protects the enterprise network from outside intrusion.

A proxy server receives a request for an Internet service, such as a Web page request, from a user. If it passes filtering requirements, the proxy server, if it is also a cache server, looks in its local cache of previously downloaded Web pages. If it locates the page, it returns it to the user without needing to forward the request to the Internet. If the page is not cached, the proxy server, acting as a client for the user, uses one of its own IP addresses to request the page from the server out on the Internet. When the page returns, the proxy server matches it to the original request and forwards it to the user. Proxies mask the return address of the requesting computer, providing anonymity for users.

An advantage of a proxy server is that its cache can serve all users. If one or more Internet sites are frequently requested, these are likely cached by the proxy, which reduces user response time. A proxy can also log transactions between it and its user group. In addition, some proxies run virus detection programs on incoming packets.

The functions of proxy, firewall, and caching can be in separate server programs or incorporated into a single package. Different server programs can be in different computers. For example, a proxy server may be located in the same machine with a firewall server or it may be on a separate server and forward requests through the firewall.

With an understanding of electronic commerce components, performance enhancements, network configurations, and security software, we turn our attention to the heart of any compelling solution, the commerce server software.

9

Selecting the Right Commerce Package

Commerce packages are the technology enablers of Web sites, often forming the backbone of the Web site's technology. These predeveloped commerce applications provide the tools and technologies needed to quickly build a site from the ground up within six to twelve weeks. Yes, e-commerce entrepreneurs still need to do the planning, understand their business, build their teams, and so on. However, these commerce packages contain a boatload of features and functions built directly into the application. The benefit? Having core Web site features such as management tools, the ability to categorize products, credit card payment systems, auction technology, and shopping cart technology built directly into a prepackaged application allows you, the entrepreneur, to spend more time researching your customers and customizing your site to suit their tastes. It gives you a competitive advantage over those who, instead, would spend months of precious development time trying to build features and functions that are expected from the consumer's perspective.

The purpose of buying versus building is time to market. Often companies will attempt to build the solution from the ground up. That's great if you have the time. However, in the e-commerce world, you must be out on the market first, with a Web site that is easy to use that provides immediate value to your customer. Most e-commerce entrepreneurs will choose a combination of e-commerce packages to empower their Web site and provide up to 80 percent of the features required for the site to be up, efficient, and valuable in the least amount of time. For example, a company may select a Microsoft E-Commerce Package and then include other vendor technologies in order to provide personalization for site visitors, inventory management, and tools for content creation and management.

Selecting vendors is essentially a three-step process. Step one, which you should have already have completed by this stage, is to have evaluated your business processes and how your company will fit into the electronic marketplace. Step two is to consider which applications will best suit your needs. Answering the following questions should prove helpful:

1. What products and services are you providing on the Internet?
2. What are your requirements?
3. What does your current technology landscape look like?
4. When do you need your site up and running?
5. Who are your competitors?

After you answer those questions, step three is to evaluate vendors who provide workable solutions based upon their quality in five areas:

1. Dominant features.
2. Commerce server management tools.
3. Integration tools.
4. Vendor and third-party plug-ins.
5. Hosting services.

Determining Your Needs

1. What Products and Services Are You Providing on the Internet?

Your answer to this question should include the essential elements that define your Web site, including your value proposition to the customer, your customers' incentives for becoming return customers, and your Web site's appeal to online shoppers.

For example, your commerce site provides toys at the lowest price on the Internet with great customer service. In this example, selling products that are shipped to customers will be straightforward, because most packages support this feature. The second key is that you want to provide great customer service. Customer service features will vary from application to application. So the key will be to define the requirements statement to include those tools needed to provide great customer service such as Question and Answer databases, allowing consumers to include product comparison, guiding your customer to the products of interest by including a Q&A database that leads you customer to a product. Other options may include online chat rooms, extensive product information, and clearly stated policies and procedures on returns, delivery information, and product tracking. You may ask yourself, then, how are they making money? Advertising! This means that they will need to have the best commerce analyses and marketing tools available so they can provide target advertising for each customer hitting the site.

Currently, Juanita Ellis is working on a project that provides a digital marketplace for designing boxes and sending out RFPs to participating suppliers. Their value proposition is that they streamline the entire process of designing boxes by providing a collaborative work space for users to download CAD drawings, change the dimensions of the boxes, and add designs on the boxes and other necessary features for package design. After the businesses have designed the boxes, they will send out RFPs to multiple suppliers. At that point, shipping dates, costs, and the like, will be negotiated between the buyer and seller. In this scenario, there are two main components. The first is the collaboration piece, which provides the real value for businesses because they can reduce the time and costs of using traditional methods of designing. This part of the site is critical and will need to seamlessly allow business groups to exchange ideas and designs online. The second component is the exchange piece, which allows the businesses to send out RFPs to suppliers. This is a great part to the puzzle, however, it is not what will bring the consumer the most value.

Other sites also may make their revenue from selling products online. However, to make their sites unique, they are also providing extensive content that is customized to shoppers accessing the site. So, when selecting a vendor solution, content management and aggregation will become critical to the success of the site. Several vendors provide these types of features.

2. What Are Your Requirements?

Once you have decided what technologies your Web site will require to function optimally, the vendor selection process becomes much simpler. For example, if your requirements consist of providing features such as online chat, product comparisons, content aggregation, and extensive personalization, you will be able to compare each package to find the one that most closely fits your needs. However, you will to need to customize. The goal here is to find the closest fit. Even with a commerce package, you will be lucky if it provides a 60 to 70 percent solution to your needs. However, that 60 to 70 percent may save your organization a lot of time and costs in the long run. With today's consulting costs of $150 to $200 per hour, using a team of six or seven developers for three months could end up costing more that the cost of purchasing the software you need and spending the dollars instead on making your site unique. Why spend the extra time and money on reinventing the wheel?

Your requirements statement will include, but is not limited to, answers to the following questions:

- What is required to sell products, content, and/or services?
- What is required for marketing?
- What is required for customer service?
- What is required to provided customization and personalization?
- What is the required content, and how will I aggregate or publish the content?
- What is required for distribution?
- What languages will I need to incorporate?

3. What Does Your Current Technology Landscape Look Like?

Review your current business process, and start mapping the current architecture and systems. The goal is to document all the existing systems that are part of the current architecture. Once you know the current architecture, you can determine what additional systems are needed and what commerce software to select.

Some of the major systems to document include network operating systems, Web server software, relational databases, transaction-based systems, ERP systems, EDI systems, and all hardware capacities and platforms. In addition, if you are integrating your systems with those of other businesses, document their systems that you need to integrate. The resulting documentation should include a high-level architecture diagram and a description of the data required from each system. Remember that your initial work requires that you capture your current architecture.

4. When Do You Need Your Site Up and Running?

In the past year, companies have discovered the importance of the first-mover advantage. Gone are the days when companies spent months and months on planning and months and months on deployment. Today's corporations and start-ups want to hit the market with a bang as quickly as possible. So it is critical that companies break from traditional methods of software selection and deployment and that they move toward getting to market as quickly as possible. In one scenario, Juanita Ellis was with a company that sought to get its first launch up and running within eight weeks. The company evaluated several consulting companies as well as different technologies. In the end, they took a two-phase approach. Company leaders selected one vendor solution for phase I, and, while building the phase I site, they selected other vendor solutions that would best fit the environment. Why? They decided that the company had to get to market within a short time frame in order to implement the other solutions. Otherwise, full deployment would have taken close to four months. An important consideration was that the first vendor package used cost just $20,000 for phase I and phase II, while the software they would later decide upon cost about $500,000. So if company leaders decided they needed to scrap the first vendor package, they had not suffered a substantial or costly setback or taken that high a risk.

5. Who Are Your Competitors?

Why would you want this information for selecting a vendor solution? Well, you want to find out what capabilities are being offered at you competitors' sites in order to ensure that your solution will be designed to provide unique capabilities. You must be one step ahead of your competitors.

Evaluate Vendor Solutions

Once these questions are completed, the vendor selection process begins. There are hundreds of commerce vendors on the market. These solutions range in cost from free (see those at www.yahoo.com) to millions of dollars. So how do you decide which solution in which price range will be most effective and economical for your site? This will depend on several factors ranging from the features and functionality required for supporting your site to the integration tools available to tie into your current infrastructure. There is a catch. You may find a great commerce solution with all of the bells and whistles, but the cost may be out of range. So, let's take a look at a range of high-end and low-end solutions.

BroadVision

BroadVision (www.broadvision.com) provides robust business-to-business as well as business-to-consumer solutions. Their solutions focus on several industries as well as providing an e-commerce storefront package. The primary solutions support retail distribution, financial services, technology/manufacturing, telecommunications, and the travel industry. The key to BroadVision's success is the features available for personalizing site content to businesses and consumers accessing the site. For example, one of the main reasons Boeing selected BroadVision's e-business platform to power myboeingfleet.com is for the capability to tailor content.

Banks such as Argentaria Bank Group and Banco Santander are using BroadVision's financial services component to provide customers with banking services online. Customers such as Audible and Blackwells have implemented retail distribution service for selling goods and services to online customers. BroadVision's products tend to be used by larger corporations that are moving toward the e-commerce arena. They also have customers worldwide.

BroadVision provides applications for several market areas. Each application is geared to a specific market. This provides great flexibility in that you need to purchase only those components pertain to your commerce site. Table 9.1 lists some of these applications.

Table 9.1   BroadVision Applications

One-to-One Billing	An e-bill delivery and payment solution that enables revenue enhancement based on one-to-one relationship marketing through customers' bills. It can also reduce the costs associated with both customer care and paper-based billing.
One-to-One Business Commerce	A complete e-commerce solution that includes all the functionality needed to establish and maintain a business-to-business e-commerce site. It fulfills seven fundamental requirements: complete functionality, scalability, manageability, relationship management, business system integration, full transaction handling, and ease of use.
One-to-One Enterprise	This is the foundation for all other modules. Its components provide individualized, real-time targeting of Web interactions through customized business rules, profiling, clickstream analysis, and tools for business users. It even supports wireless standards such as WAP.b.
One-to-One Financial	Provides several components that include Real Time Quotes, quote services that enable institutions to offer personalized stock tickers to their customers for tracking their individual stock portfolio on a real-time basis, customer service functions for managing users, alerts on quotes, and bank balances.

(continued)

Table 9.1   BroadVision Applications (*continued*)

One-to-One Instant Publisher	Provides Web-based tools that allow even nontechnical users throughout an organization, in multiple geographic locations, working on multiple computer platforms, to publish content in multiple languages, personalized to reach the appropriate recipients. Support is provided for content created with any tool, including HTML editors, Microsoft Office products, Lotus Notes, and Documentum.
BroadVision MarketMaker	A complete, out-of-the-box application for creating and managing online B2B marketplaces.
BroadVision Procurement	Automates the purchase process, from generating the requisition to generating the purchase order. It provides real-time order status and easy access to requisition and order history and is fully integrated with and certified by CommerceOne MarketSite.
One-to-One Publishing	A powerful content management system that enables companies to manage and deliver targeted, timely, personalized, and media-aware electronic content.
One-to-One Retail Commerce	Offers business-to-consumer functionality, including personalized contents, advertising, cross-selling, and other features required to provide B2C functionality.

OpenMarket

OpenMarket is another dominant vendor of e-commerce solutions. OpenMarket's customers include Lycos, America Online, Consumers Union, Business Week, Standard & Poors, *USA Today*, Playboy Online, RealNetworks, Acer, Ingram Micro, Milacron, and Cablevision.

OpenMarket is also the leader in the commerce service provider market with a worldwide infrastructure of telecommunications providers, banks, Internet service providers, and portals. Some of these companies include AT&T, Barclays Bank, First Union National Bank, France Telecom, Hiway, Netcom, NTT, Sage Networks, SBC Communications, Swiss PTT, Telecom Italia, and Telstra. OpenMarket has also provided several solutions for smaller businesses to get online at very little cost. OpenMarket's comprehensive Web site (www.openmarket.com) details additional customer solutions, ISVs, and hosting partners.

OpenMarket provides solutions for content management and commerce management. This vendor also provides and suite of applications, each providing specific functionality to the site. Table 9.2 includes some of these applications.

Table 9.2   OpenMarket Applications

OpenMarket's Content Server	Provides for rapid deployment of systems that enable multitargeted delivery of content. As a result, businesses can publish a single asset to multiple Web sites, cell phones, or other Internet appliances.
Transact	Allows merchants to manage orders from the initial purchase of an item to the updating of internal inventory systems. Transact provides capabilities to capture, store, and integrate with existing systems and to process various forms of purchase payment. In addition, the software allows you to fulfill orders and send that information immediately to your customer service department. Other features include online customer authentication and authorization, automated tax and shipping calculations, and online customer service. Also supports partial shipments and backorders.

(continued)

Table 9.2   OpenMarket Applications (*continued*)

Content Centre	A browser-based application that enables nontechnical users to create, manage, manipulate, and deliver content to multiple targets, including multiple Web sites, e-mail, and wireless devices.
Personalization Centre	Enables businesses to implement site personalization, including personalization based on concept mapping, creating and storing visitor profiles, and creating and managing the segmentation of visitors.
Catalog Centre	Helps vendors create and maintain an online catalog.
ShopSite	Allows medium-to-small merchants to set up an online store in as little as fifteen minutes. The store creation wizards provide all of the features for a fully fuctional e-commerce solution, including an "on sale" module, product up-selling and cross-selling, associates tracking, order management, and processing.

Ariba and CommerceOne

Ariba and CommerceOne are two competing vendors in the procurement and exchange market. These vendors provide very similar solutions for large organizations or companies participating or hosting an exchange. Both companies customers reads like a "Who's Who" list from several of the major markets.

Ariba customers range from large manufacturing companies such as Volkswagen and Honda Motors to energy companies such as Chevron and Texaco. Ariba has been the leader in procurement and market exchanges. Their solutions consists of the applications listed in Table 9.3.

Table 9.3   Ariba Applications

Ariba Buyer	Ariba's answer to the procurement needs of today's global enterprises. This single solution can be used to buy both direct and indirect goods, acquire services, track travel costs and other expenses, and analyze strategic spending. The application facilitates the buying process and claims to reduce costs, eliminate inefficiencies, and speed up the transaction flow from end to end.
Ariba Marketplace	The industry's leading marketplace application: a flexible, rapid-deployment solution for market makers of all kinds. Suitable for all types of B2B exchanges—from horizontal procurement marketplaces to highly specialized vertical marketplaces and commodity exchanges—the application bridges the gap between traditional buy-side and sell-side solutions. An all-in-one solution developed for market makers to efficiently deploy effective online trading communities.
Ariba Dynamic Trade	A highly configurable, fully integrated auction and exchange application that delivers the full range of dynamic pricing mechanisms—auction, bid/ask exchange, and reverse auction—in a flexible architecture. Supports B2B trading situations including multitier channel relationships, approved vendor lists, price and bidder confidentiality, time sensitivity, and differing quality or support levels.
Ariba Commerce Services Network	Tightly integrated with the Ariba B2B Commerce Platform, the Commerce Services Network provides sourcing, liquidation, supplier content and

(continued)

Table 9.3   Ariba Applications (*continued*)

directory services, secure transaction routing, multiprotocol exchange of transaction information, payment services, and logistics. Ariba's leading set of network-based commerce services helps businesses streamline their supply chain and seize new revenue opportunities. Offered both directly and through partnerships with other leading service providers, these services allow businesses to take maximum advantage of online B2B trade.

CommerceOne provides similar solutions to those offered by Ariba. Some of their customers include Lockheed Martin, Boeing, Compaq, and General Motors. The solutions offered by CommerceOne provide both procurement and exchange capability. These solutions are listed in Table 9.4.

For smaller businesses going online, some of the solutions above may fit the bill, although they are very expensive to purchase. Commerce service providers or application service providers provide a viable alternative to purchasing the package directly. With these providers, merchants can host their application and rent these applications on a monthly or per-transaction basis. However, if this is still out of reach, there are other commerce packages aimed at the smaller businesses. Most of these applications are geared toward business-to-consumer online sales or pure auctioning sites. Let's take a look at a few of these solutions.

ICat

iCat Commerce Online (www.iCat.com) is a storefront creation service geared for small businesses that are new to e-commerce and need a fast solution within considerable budget constraints. iCat Commerce features include extensive sales reporting and real-time credit card authorization. At the time of publication, iCat charged a one-time fee of $249, then a $49.95 monthly charge to host up to fifty items. With ten or fewer items, there was no monthly charge.

Table 9.4   CommerceOne Applications

BuySite 6.0 Enterprise Edition	Comprehensive procurement functionality that addresses the entire purchasing cycle from requisitioning to payment. Features include managed content via a buyer-managed, multisupplier catalog; requests for ordering off-catalog of one-time purchases; and services for ordering or configuring items from a supplier Web site.
MarketSite Global Trading Portal	The U.S. portal for exchanging goods and services among businesses worldwide. This open portal integrates with a wide variety of buying and selling applications, including CommerceOne BuySite, and it connects to the CommerceOne Global Trading Web.
Global Trading Web	CommerceOne's Global Trading Web (GTW) is the world's largest business-to-business trading community. Comprised of many interoperating portals, the Global Trading Web provides economies of scale for buying organizations, suppliers, and service providers worldwide. Each portal is independently owned and operated by a leading market maker in a region or industry. The result is the largest choice of suppliers, prices, and selection for buyers; the largest market for goods and services for suppliers; and new efficiencies and opportunities for market makers.

MerchandiZer

MerchandiZer (www.merchandiZer.com) is a simple, low-priced solution for small-business owners. For an initial setup fee of $149 and monthly hosting fees starting at $79.95, MerchandiZer delivers support for unlimited items and storage space. MerchandiZer is an online-only solution that allows merchants to build an online store with an Internet connection and a Web browser. The Web site's setup is handled through MerchandiZer's manager utility, where merchants select from a number of prebuilt themes and templates for their storefronts. They can also add custom graphics to differentiate their Web site.

IBM HomePage Creator

IBM HomePage Creator for e-business is a new entry in the small business e-commerce field, and it boasts a top-notch storefront creation tool. With a one-time setup fee of $24.95 and a monthly fee of $24.95, the basic plan offers merchants five pages, twelve catalog items, 3MB of storage, and 40MB of bandwidth. Furthermore, it promises to automatically list storefronts with twenty search engines.

Although a few solutions have been covered here, more complete lists are available at the following Web sites:

http://ecommerce.internet.com/resources/vault/commserver/
http://www.ecommercevendors.com/

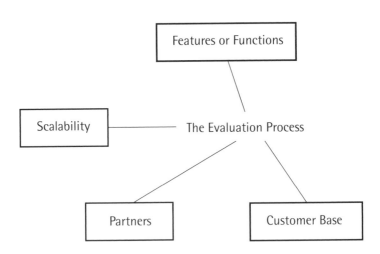

Figure 9.1 The evaluation process.

Regardless of the package, you will want to evaluate the following on each commerce vendor solution available. This evaluation process should consist of the items shown in Figure 9.1 (these may vary based on the business).

Many e-commerce software solutions include standard store templates that can be readily customized or storefront creation wizards that will walk you through the construction phases of building a store. Several vendors have sample stores that you can use as a basis for the site. The selected template serves as a model of your eventual store and represents the hub of your e-commerce site.

After you walk through the wizard screens, the site still should require modification. It's important to differentiate your site from others in branding and online features. The most common question about these wizards is whether people really use them. The answer is yes, if they are creating a storefront to sell products or services. The rationale is that these wizards contain the basic information you need to create the site, such as the storefront home page, product information, tax and shipping information, credit card information, and design and layout. These wizards also allow you to easily integrate third-party credit card verification systems (such as CyberCash or VeriFone), Taxware International services, UPS tracking capabilities, and inventory tracking systems. The integration of these third-party components represents the most compelling reason to at least consider the wizards.

The basic principle behind wizards is simple: Why reinvent the wheel? It is much easier to create a store using these tools than it is to construct a store from scratch. If you build your solution from the ground floor up, it may take several developers several months to reach the same outcome offered by the wizard-based solution. Put your efforts where your expertise is most developed. Where it is less developed, harness the expertise of others. In the field of e-commerce, time is your primary opponent. Do not spend time learning a skill that you can rent instead.

When selecting a commerce solution, it is important to review the built-in templates or wizards that are part of the product. Each vendor has solutions that vary slightly in their out-of-the-box capabilities. By purchasing a package that meets your business needs, you can reduce the amount of time and resources required to customize your solution. You must compare the cost of a standard software solution with that of a custom solution designed and developed by internal or external resources.

Most commerce vendors provide business-to-consumer and business-to-business capabilities. However, as the e-commerce field is gaining greater grounds, vendors are focusing on key areas of business.

Platforms Supported and Integration Tools

Each commerce vendor solution will support specific network operating systems, Web servers, and out-of-the-box, back-end integration tools that allow you to connect your databases to your e-commerce solution. To make a reasonable decision when selecting one vendor's product over another, you need to identify and define exactly what systems you have in house so you can determine if the solution will easily fit into the environment. Based on your business needs across the various components of your infrastructure, ask the following questions:

1. What network operating system (NOS) is required?
2. Which Web server is required to run the chosen commerce software? Some vendors allow you to run their commerce software on several types of Web servers available on the market.
3. Does the vendor provide plug-in tools to the relational databases currently deployed at my or my business partner's company? Relational database systems supported by these vendors include Oracle, SQL, and DB2.
4. Does the solution have the capability to integrate into my transaction-based systems (often, CICS and IMS)?
5. Are there tools that provide connectivity to the enterprise resource planning (ERP) systems currently used by my organization? ERP systems include SAP, PeopleSoft, Baan, and J. D. Edwards.
6. If you employ electronic data interchange (EDI), what connectivity is provided by the vendor solution?

Also, determine if plug-ins are available to any other third-party software solution you may currently employ. These software systems include document management software, image management software, and accounting packages.

Ease of Modifications and Extensions

It is important that the solution selected be fairly flexible as far as making modifications to work flow, integrating other applications, and third-party plug-ins. Some commerce vendors provide solutions that are easy to modify and include third-party applications integration. These third-party plug-ins include integration of credit card payment systems, customer relationship management systems, collaboration software, and tax and accounting and marketing and analysis applications. Third-party vendors are building plug-ins for most of the larger commerce vendors. Because a commerce package can provide between 40 percent and 60 percent of a solution, it is important that the package you select allows you to easily add features and functionality to your site. Your must be able to modify and add to the total solution as technology changes and your business grows without the time- and money-intensive process of overhauling it. Based on your business needs, find out if the commerce solution supports the plug-ins required for your site.

Companies listed as supporting the commerce application are usually listed as independent software vendors or technology partners. You can find this listing directly at each commerce vendor's Web page. For example, at BroadVision's technology partner web page, some of the following partners are listed: CyberCash, CyberSource, Epiphany, Interwoven, Macromedia, Moai, Net Perceptions, Oracle, PaylinX, Sybase, Taxware International, and VeriFone. Ariba also has a technology partners Web page that list vendors such as BEA Systems, Extricity Software, i2 Technologies Inc., J. D. Edwards, Mercury Interactive, and Siebel Systems, Inc.

Companies such as IBM and Microsoft provide their own plug-ins for their e-commerce software packages. For example, both IBM and Microsoft sell an auction technology plug-in. Other commerce vendors provide plug-ins for package tracking through UPS or FedEx, additional management tools, and development tools. This information is usually provided at the vendor's Web site. Because these vendors continue to improve their products via plug-ins, it proves very advantageous for you to keep updated. Visit vendor sites at least every couple of weeks to determine what additional tools have been developed to integrate into your e-commerce solution. Remember that it is much easier and faster to deploy a plug-in than to create the same functionality.

Scalability

Depending on the applications running on the server and the number of concurrent users hitting the system, the number of servers required and the hardware and software requirements will vary. When selecting a software vendor solution, it is critical that the application architecture provides features that manage the load and provide scalability. When selecting a solution, ask the vendor questions that include:

- What tools do you provide for load balancing?
- Does the server provide caching features?
- Do you have analysis tools that will allow us to test the server performance based on applications running on the server and simulate concurrent users?
- Is it easy to distribute the application so we can run different processes on different servers in order to balance the load?
- Has the application best tested in a production environment? What was the result?

Partners

Partnerships are essential in e-commerce. Your partners will help you design and deploy a solution and host your commerce application. Partnerships fall into three main areas:

1. Consulting firms
2. Technology
3. Hosting

Certain consulting firms have partnered with each commerce vendor in order to help customers design and develop the site. Because each technology and solution will vary, it is important that the consulting firm you select has worked with the products you are evaluating and have deployed solutions in similar industries. By selecting an experienced consulting firm, you reduce the risk of designing a site that will fail. In addition, consulting firms have developed their own best practices that will provide a methodology and foundation for developing and deploying your site. Think about it. Consulting firms have been involved with a number of customers in your type of business and will be able to take their experience to your project. Yes, there is the risk that they may take one of your ideas and bring it to a competitor. If you've selected the right firm, however, by the time one of your ideas reaches a competitor, your solution is up and running effectively on the Internet. This leaves you free to think actively about developing a newer, heightened function that will leave your customers saying, "Wow!"

We discussed technology partners in the previous sections. These technology partners will provide solutions that have been implemented as part of the commerce solution. You will want to find out how to integrate each of these solutions into your commerce site prior to making the decision.

Hosting companies that participate in these partner programs have experience managing and implementing the commerce vendor solution. Not every hosting company has experience in each application; it is important that you select the hosting company that does. In addition, we strongly recommend getting references from each partner to discern the reliability, performance, and support of each hosting company. Although they may be listed as a partner, you cannot always assume that they are the best fit for you company. The best way to find out is to directly ask the businesses that have experience with the hosting partner about their experiences with that partner.

Customer Base

Who are the vendors' customers, and how are they using the proposed vendor solutions?

When choosing your vendor solutions, you *must* ask this question. Your greatest fear at this point should be choosing a vendor solution that will use your Web site as a testing ground. Although cost incentives may exist, merchants cannot have their Web solutions serving as a learning opportunity for a vendor. If a vendor has successfully deployed their solution, your confidence can be buoyed. Otherwise, heed your natural skepticism. If the vendor provides you with a significant list of customers using the same solutions you are interested in, then ask the commerce vendor for customer references from this list. Your goal is to determine if their solution met other customers' expectations, what difficulties other customers experienced, and whether these customers would recommend the product. Furthermore, by spending the time required to review the vendors' customers, you can gauge their overall experience level in different industries, including content providers, manufacturing, travel, and entertainment.

A Final Reminder before You Buy

Before you select a vendor solution, remember that selecting the best possible vendor solution for your venture relies absolutely on how you have determined your company will fit into the e-commerce marketplace. That is, you must enter the vendor selection process with a well-defined vision statement, set of goals, and strategies and have refined any business processes. If you understand fully how you intend to exploit this compelling opportunity, choosing a solution becomes painless and profitable. If not, the sea of vendor solutions is vast, rough, and practically unnavigable.

Reading through the chapter, you may have noticed that some vendors provide solutions that cater to selling to other businesses rather than selling to consumers. Other software solutions include features that support specific certain industries, such as banking, manufacturing, and travel. Still other software vendors provide solutions geared toward selling digital content versus selling products that are physically shipped to customers.

Once you know where to move your business in the electronic commerce marketplace, judge each vendor based on the five basic guidelines outlined in the chapter:

1. Dominant features
2. Commerce server management tools
3. Integration tools
4. Vendor and third-party plug-ins
5. Hosting services

Then choose and deploy.

10

Secure Your Investment: Security Threats and Solutions

In the previous chapter, we addressed security issues regarding online payment systems. Although they are most visible to the customer, those risks encountered in transactions are just the tip of the e-security iceberg. The Internet can provide as many security risks as opportunities for a company—and some would number the opportunities online "infinite."

Security issues—hackers, viruses, and the like—are frightening issues for companies with any level of Internet connectivity. Often companies address these issues by focusing only on the hardware and software needed to keep hackers and viruses outside their intranet systems. Hardware and software considerations represent only one layer of security. Implementing solid security measures also means establishing policies and procedures that require users and network administrators to take precautions.

Security threats have steadily increased from both outside and inside corporate boundaries. Traditionally, 80 percent of security threats have been generated from within a corporation. These threats typically include employees within a company gaining access to company systems through loopholes in security and by access to employee passwords.

These percentages are changing due to the growth of the Internet and corporate extranets. Intruders from outside your corporate boundaries now represent a growing concern.

Research conducted in early 1999 by the Computer Security Institute with the participation of the San Francisco Federal Bureau of Investigation's Computer Intrusion Squad revealed that both internal and external security threats are rising. For companies that acknowledged unauthorized use in 1999, 43 percent reported between one and five incidents originating outside the organization, and 37 percent reported between one and five incidents originating inside the organization. This report was based on responses from 521 security practitioners in American corporations, government agencies, financial institutions, and universities.

The report offers some other striking considerations:

- Companies that reported their Internet connection as a frequent point of attack rose for the third consecutive year from 37 percent of respondents in 1996 to 57 percent in 1999.
- Unauthorized access by insiders rose for the third consecutive year with 55 percent of respondents reporting incidents.

With the number of security threats and their severity increasing, e-commerce entrepreneurs must take proactive roles in reducing their vulnerability. However, it is important not to fall into "security paranoia" by implementing such high security measures that employees and customers are prevented from communicating effectively.

This chapter focuses on methods to reduce security risks in areas most vulnerable to hackers, viruses, and other forms of network intrusion. Specifically, we explore the following security issues: network threats, security policies and procedures, firewalls and proxy servers, encryption, digital signatures, authentication and access control, virus prevention, and auditing your network.

Keep in mind when developing a security strategy that you must balance the need for communication and information transfer between your company and customers, vendors, suppliers, and the public, with the potential for harm that this access creates. Understanding security risks, especially those associated with electronic transactions, helps your company design and architect a secure infrastructure.

Network Threats

Hackers continue to invent new techniques for accessing internal networks and personal computers. One Web site, www.alt2600.com, provides information on the latest sites attacked by hackers and links to tools that they use. The sections that follow provide information on some of the ways intruders compromise security on corporate networks and personal computers.

Port Entrance

Several hacker tools, available free on the Internet, scan a system's open ports in an effort to gain entry. For example, Portscan 1.2 is a utility that allows the user to scan ports on any target system. The user specifies the target IP address. The program then scans all ports between port 1 and port 65536. The resulting information can be used to find loopholes in a security setup. In addition, the program identifies FTP and WWW services on any assigned port. A similar tool called IP-Prober allows open ports to be scanned in order to gain entrance to the network.

To prohibit intruders from entering open ports, disable all ports and enable only those that are required to gain access to the network. Once you enable the selected ports, use firewall security to prevent unauthorized access. The firewall section of this chapter describes port security in more detail.

Password Cracking

Password cracking represents another method hackers use to enter your network. Several cracking tools, available throughout the Internet, generate and test a series of potential letter and number combinations until they determine the correct password.

To prevent password cracking, follow these basic guidelines. Require users to create and use passwords of eight or more characters that combine alphanumeric elements. In addition, many system administrators regularly run a tool called Crack. This tool tests users' passwords to determine if they are easy to crack. This tool is available for download at www.cert.org.

E-Mail Spoofing

E-mail spoofing results when a user receives e-mail that appears to have originated from one person but was actually sent by another person. The goal of this spoofing is to trick the user into divulging information or replying with information that is confidential.

One method of preventing spoofing is to use electronic signatures to exchange authenticated e-mail messages. Electronic signatures, also known as digital signatures, provide a mechanism to ensure that messages have not been altered during transmission and that the messages are from the person listed as the sender. E-mail offers options to send messages that incorporate digital signatures.

Virus Intrusion

Virus intrusion represents a constant threat. Virus exposure results from receiving e-mail, sending documents over the network, and even from installing software or copying files from another hard drive or diskette. In 1999, the Melissa virus, the CIH/Chernobyl virus, and the Happy99 virus made headlines, causing huge damage to personal and corporate computers around the world.

On March 27, 1999, the Melissa virus began propagating via e-mail attachments. When attachments were opened in Word 97 or Word 2000 with macros enabled, the virus infected the Normal.dot template causing documents referencing that template to become corrupt. If the infected document was opened by another user, the virus would propagate. Because the virus continued to propagate, mail servers experienced performance slowdowns.

On April 26, 1999, a virus called Chernobyl (CIH) started spreading. This virus infected executable files and then spread by executing an infected file. Eventually, when the CIH virus was activated, it began deleting information from hard drives and overwrote each system's BIOS.

On January 20, 1999, the Happy99 virus hit the Internet. The Happy99 virus is an executable file that displays fireworks and the words "Happy 99" on a user's screen. During its run cycle, the virus modifies files on the system.

In early May 2000, the ILOVEYOU virus, also coined Love Bug or Love Letter, caused unprecedented damage. The self-propagating and destructive "ILOVEYOU" virus crippled mail servers and destroyed jpeg image files at tens of thousands of networks around the world. Love Letter spawned at least twenty-five copycats with varying levels of destructiveness. Computer Economics, a research company based in Carlsbad, California, estimated that the virus and its variants caused $6.7 billion in damage. The virus was first reported in Hong Kong and spread rapidly west, infecting government and business computers.

With viruses attacking both the corporate and private world of personal computers, most computers feature virus protection software. If you do not have a version of this software, you can obtain it via the Internet from a variety of vendors. Remember to update the virus protection software to ensure you have protection against the most recent viruses. For corporate organizations, institute the appropriate policies and procedures to require users on the network to deploy and use the latest virus protection software.

Good informational sites for virus threats, research, and protection include:

- Symantec AntiVirus Research Center:
  www.symantec.com/avcenter/index.html
- IBM's AntiVirus Online: www.av.ibm.com
- Virus Bulletin: www.virusbtn.com

Other Hacker Attacks

Hackers employ a variety of other methods to attack servers around the world. This section describes a few more areas of risk. Most of these hacking methods are preventable using sound security policies and appropriate software and hardware implementations.

The "ping" command can be used to exploit your system. Several free hacking tools allow hackers to send either oversized packets or continuous packets to your server. The result is a system interruption or even a crash. For example, Ping-g is a tool used to specify an IP/Host address to continuously send a specified amount of ping packages. Another tool is the Ping of Death, used to send oversized packets to the server, resulting in a system crash.

Most operating systems now include the functionality of accommodating these larger packet sizes. In addition, if a hacker is using a tool such as Ping-g, monitor your system activity. If necessary, deter the sending system from accessing your server.

A tool called SScan performs probes against servers to identify services such as Telnet and FTP. You may be vulnerable to access by an intruder. The tool can be configured to automatically execute scripts of commands on your system, causing major problems. To prevent the SScan tool from locating vulnerabilities, use firewall filters to monitor any TCP/IP application ports and to detect unauthorized scripts running on the system.

Security Policies and Procedures

Any concerted effort to counter common network threats begins with comprehensive administrative policies and procedures. User-level security focuses on ensuring that employees are aware of the importance of security within the organization. Administrator-level security ensures that hardware, software, and user access is documented and granted according to the established guidelines. The following section describes the most critical areas to consider when defining security policies and procedures. Depending on your company's needs, some requirements may vary in how intensely they are monitored. For example, in environments of highly confidential information, more time and money resources may be expended to deploy multiple layers of firewall security instead of just a single layer.

User-Based Security

User education and security awareness programs should be the starting point for developing a secure environment with the support of your employees. Often, employees work unaware of the implications associated with the following situations.

Do you allow others to access the network with your user name and password? You may have asked, "Can I log on with your name and password to get access to the network?" The response is typically, "Sure, no problem." Actually, the problem has just started. Ensure that personnel within the company understand the importance of not allowing others to use their login name and password. If someone is logged on as another user, it becomes difficult to determine who actually caused a security breach.

Do you leave your computer on when you leave your work area? Do all of your colleagues? Most do. This situation creates a risk that most people may not even consider. It is time for lunch or a meeting, and you are working on a document. Instead of logging off the system and turning off the computer, you just walk away from it. Now consider that a virus can be deployed within 30 seconds or less. An entire system could be infiltrated or otherwise compromised during one employee's jaunt into the next office. Establish a policy that requires users to log off prior to leaving the work area. This simple action can save a company a lot of time and money.

Do you write your password on a note lying next to the computer or stored in a drawer? Our workplace has become the "login name and password" game. Each system requires multiple login names and passwords. However, writing them down on a piece of paper means not having security at all. Ensure that users do not practice the "sticky note" habit. Remind them how critical these user names and passwords are to protect their data as well as company information.

Users must understand that the information they are granted access to is secure and is confidential. Users must be informed that they are given different degrees of access to sensitive company information based solely on their user name and password.

Companies can take additional steps to help users manage confidential information. For example, when sending e-mail or documents electronically, an option is provided to include the word "Confidential" in red letters across the top. This tagging informs employees who may have access to this information that the document is of a sensitive nature and should not be sent or discussed with others in the company.

Administrator-Based Security

Administrator security policies must be established to protect company information and computer systems. These rules dictate both a protective arrangement of hardware and software and a series of regular practices to ensure an organization's technical health. Setting security policies and procedures begins on a physical level and includes operating system security, Internet security, remote access rules, virus prevention, and commerce application protection.

Physical Server Security

Physical server security is the first level of policies and procedures that must be established. Keep servers in a locked room with controlled access. Someone who can get to the server can often very easily access the information. Make sure these servers are secure, and register each person who enters and exits the server room.

Protect keyboard access with passwords. If controlled access is breached by an intruder, the security afforded by passwords may protect your company. Exercise reasonable caution by ensuring password lengths are more than eight characters and by using alphanumeric characters. Systems requiring double passwords to access information might be appropriate in some circumstances.

Establish security procedures to prevent unauthorized users from gaining access to the network. Some companies require badges and digital cards to access computer rooms. Other companies require passwords prior to allowing access to the computer center. However, based on our corporate experience, administrators have often given their access information to outsiders, such as consultants working on site. Establish and enforce policies and procedures for these situations. You do not know who may be lurking in the hallway.

Basic Operating System Security

The operating system requires its own set of security rules and regulations. Even if you have set security for the Web server through firewalls and proxy servers, access through the operating system remains a big risk. The following sections suggest a variety of security policies.

Define user, server, and group access rights to files and directories. You must determine who has read, write, delete, and modify access to directories and files on the system. It is also a great idea to set up a "no access" group. This is especially useful if someone from the company leaves the organization. You can easily and immediately place that user in this group, preventing access to system files.

Require passwords to be greater than eight characters in length. This will make it more difficult for users trying to access the system. Furthermore, require a combination of alphanumeric characters which are more difficult to crack. Another tip is to set up the system so users are required to change passwords every two to three months. Again, this provides another layer of protection.

Place system files on a separate partition. This will make it easier to manage the information. For example, place all of the system files on the C: drive, and give only administrators access to the information. Once you have done this, log all additions, deletions, and modifications to these systems files. These are the files hackers seek. Also monitor server log files for suspicious queries or unusual system access.

Limit almost all root access. Someone who has root access can perform basically any command on the system. This includes adding, deleting, and monitoring of files.

Internet and Port Security

Internet security policies and procedures include defining Internet access and ensuring port security. Define Internet (external) access for buyers, suppliers, trading partners, and customers as a whole. The best method to set these policies is to determine the main groups who require access and then determine their level of access to the information on your site. For example, you may have company A, company B, and anonymous Internet customers. Once you have set up these groups, setting up site access proves much more manageable. Imagine that company A has access to certain products you are selling at the site. When members of company A log on, they immediately are placed at the site that contains products and services specific to their company. In addition, you can provide specific vendor discounts and server access (based on user login name and password).

Port security is also critical to Internet security policies. Internet port and security configuration allows enabling or disabling port 80 for HTTP access and enabling and disabling port 443 for SSL access. When Web clients access the server, they access port 80 by default, similar to a door that allows access to the Web server. When this door is opened (enabled), they have access to the information at your store. This port should be enabled to allow customers to access the information at the site. However, there are also several other ports that provide an open door to other applications on the network. For example, 20/21 is used to access the FTP server on the network. This port should not be displayed if you do not plan on having an FTP server for your e-commerce site.

The bottom line is to allow access only to those applications, such as HTTP, that are required by your customers on the Internet.

Firewall and proxy server security adds another layer of security to your site. The firewall is used to protect your intranet or internal network from the Internet. As a complement, proxy servers limit the sites which your internal users can access on the Internet. We discuss firewall and proxy server security in more detail later in the chapter.

Security for Servers with Dial-In Access

Policies and procedures are also required when providing access to servers that have dial-in capabilities. When setting up a remote server, ensure that the server is behind the firewall, and follow strict access and operating system security procedures. Other forms of security include using dial-back and password protection for modems. Keep in mind that as soon as someone dialing into the network is online, he or she is participating in your environment as another node on the network.

Virus Protection Software Security

Policies and procedures on virus protection software must also be established for the company. For example, require that virus scanning software is installed on all computers and that updates are provided on a regular basis. Because new and updated viruses are a constant threat, a systematic approach to distributing these updates to users within the company helps protect the systems. Other measures include prohibiting users from copying/installing files or programs from home onto corporate computers. We discuss viruses in more detail later in the chapter.

Commerce Application Security

Commerce application security includes defining those users and groups who have permission to perform specific tasks on the commerce server. Ensure that there are defined policies about who has access to perform particular tasks on the commerce site. Most commerce server software packages provide the capability to create user groups and assign access rights to each level of your storefront. These rights define who has the ability to add, modify, or delete product information, approve customer orders; and modify site design, and they include administrative rights to perform all tasks on the Web site.

Separating Your Net from the Net: Firewalls and Proxy Servers

Firewalls and proxy servers provide security between your intranet and extranet and/or the Internet. Organizations install firewall software to protect their intranet from unauthorized external access. A firewall works at the hardware or software level to control access to the internal system. Firewalls implement such security measures as packet filtering, application-level gateways, and circuit gateways, and they are often used in conjunction with a proxy server. A proxy server mediates between client requests in your company, such as HTTP requests from a browser, and servers outside your firewall. Proxies mask the return address of the requesting computer, providing secure anonymity for users and denying any potential targets. Some proxies run virus detection programs on incoming packets. The following sections describe firewalls and proxy servers.

Firewalls

In simple terms, a firewall consists of a server and two network interface cards. The external network card communicates with the Internet or extranet, and the internal network card communicates with the intranet or internal network. In Figure 10.1, the firewall is placed between the internal servers and the external servers. This firewall design provides protection against external users accessing the internal network.

Firewalls provide a method to filter information that is transmitted over the network. This method is sometimes called a datagram or packet. A packet-filtering firewall examines where the packet is going and what type of information it contains. It then checks whether the network allows delivery to that destination and allows that type of information to enter or exit your network. If the packet passes these tests, it is allowed to continue to its destination. Typically, packet filtering is implemented with router software.

Packet filtering provides a way to filter IP addresses based on IP addresses and ports. For example, you could allow only known IP addresses access to the network. On the other side, you can deny access to all unknown or unreliable IP addresses (including competitors).

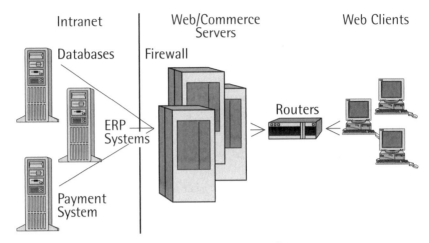

Figure 10.1 Firewalls are an essential measure for protection.

Another example of firewall security involves denying all outsiders access to port 80, resulting in no one from outside the organization accessing the HTTP server. If you decide not to implement this type of security, consider an alternative. The same result could be achieved by assigning another port to the HTTP server. Those intruders who attempt to violate the system security via port 80 would be unsuccessful.

What firewalls can do for your network:

- Prevent unauthorized logins.
- Block outside traffic from inside your network.
- Filter unwanted network traffic.
- Provide security and access logs.
- Act as an effective tracing or monitoring tool.
- Enforce security policies.

What firewalls *cannot* do for your network:

- *Cannot* protect against traffic that does not flow through the firewall.
- *Cannot* protect against bad, unrealistic, or nonexistent security policies.
- *Cannot* protect against viruses.

- *Cannot* protect against data-driven attacks, such as e-mail, FTP, or copied data, executed on the host.
- *Cannot* protect against data theft by authorized users, such as industrial spies or disgruntled employees.

When planning your firewall architecture, the first step is to define firewall policies. The planning process is often the most time consuming and cost intensive. At this point, you must decide who has access, what servers they have access to, and what elements to monitor on the system. Based on security needs within the company, select the vendor software solution that addresses the security requirements defined for your corporation's architecture. Table 10.1 lists a variety of major firewall vendors, their products, and contact information.

Table 10.1   Top Firewall Vendors

Company Name	Product Name	Web Site Address
Ascend	Ascend Secure Access Firewall	www.ascend.com
Cisco	Cisco IOS Firewall	www.cisco.com
Sterling Commerce	CONNECT Firewall	www.sterling.com
CyberGuard Corp.	CyberGuard	www.cybg.com
LanOptics	Guardian	www.lanoptics.com
Microsoft	MS Proxy Server 2.0	www.microsoft.com
Check Point Software Technologies	Fire Wall	www.checkpoint.com

Proxy Servers

A proxy server functions as an intermediary between a secure network and a nonsecure network. Most firewall products offer proxy server capabilities. With a proxy server, you can monitor information such as who has access to particular sites on the Internet. If necessary, you can prevent users within the company from accessing specific servers on the Internet. For example, the user can access a site if the Web server name is www.microsoft.com. However, if the user then tries to

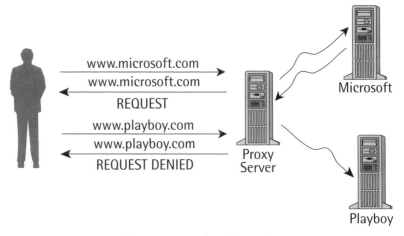

Figure 10.2 A proxy server allows you to monitor information.

access www.playboy.com, the proxy server has been configured not to allow that access. This situation is illustrated in Figure 10.2.

Proxy servers also include a caching service for Web pages. The next time a user accesses a previously cached page, response time improves because the server does not have to access the Internet to reproduce the page. The process of caching often dramatically improves response times. A proxy server receives a request for an Internet service, such as a Web page request from a user. The proxy server passes its IP filtering requirements, or firewall, and looks in its local cache of previously downloaded Web pages, or cache services. If it finds the page, it returns it to the user without needing to forward the request to the Internet. If the page is not in the local cache, the proxy server uses one of its own IP addresses to request the page from the server out on the Internet. After the page is received and filtered, the proxy server forwards the page to the user.

When setting up a proxy server, configure it as application-based or generic. Depending on how you plan to monitor user access to the network, you can select either of these two options.

An application proxy uses the same method as a packet filter. It examines where the packet is going and the type of information it contains. However, an application proxy does not let the packet continue to its destination. The application proxy delivers the packet for you.

An application proxy firewall is a server program that understands the type of information that is being transmitted, for example, HTTP or FTP. Application proxies control the flow of information between internal and external clients and servers. This option may be used if you want to control access to FTP sites but not specific Web servers. Web browsers have built-in settings to specify proxy server configuration. For example, the browser contains a setting to require a proxy prior to accessing the Internet, FTP, Gopher, or Web servers.

A generic application proxy is similar to an application proxy except that it does not need to understand the type of information that is being transmitted. For example, SOCKS servers can function as this type of proxy. This method may be used if you want to monitor traffic from your intranet to the Internet.

Firewall and proxy server planning and implementation are critical to securing your corporate information and helping to prevent unauthorized user access. Firewalls are used primarily to control traffic from the extranet and/or Internet to your corporate information systems. Proxy servers, on the other hand, are used to control sites that employees can access from the internal network to the Internet. In the following sections we discuss other areas that are required to prevent corporate information and data from falling prey to a hacker.

Encryption

Encryption is the transformation of data to a scrambled form that cannot be read or interpreted without the appropriate translation key. Encryption and decryption require the use of common information, such as a key or key algorithms, to scramble and unscramble data. Encryption and decryption are crucial online, because cryptography is the only practical method of protecting information transmitted electronically.

Consider that contracts, documents, money orders, and other legal information travel across the Internet. With the threat of hackers, wiretaps, competitors, and dishonest insiders, a method must exist to scramble data while stored on the system or during electronic transmission. Understand the following few data encryption methods in order to understand how to use this technology to empower your own business: symmetric key encryption, asymmetric key encryption, and Pretty Good Privacy freeware.

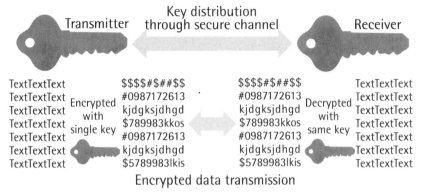

Figure 10.3 Symmetric key encryption.

Symmetric Key Encryption

Symmetric or single-key encryption features a single key to encrypt and decrypt information. This technology works well when sender and receiver can arrange ahead of time to exchange the key. In most business cases, such exchanges are cumbersome, especially among multiple parties. Figure 10.3 illustrates symmetric key encryption.

Asymmetric Key Encryption

Asymmetric or public-key encryption requires two complementary keys to encrypt and decrypt information: a private and a public key. The public key, freely distributed, is only used for encryption and is not able to decrypt any encrypted documents or files. Only the complementary private key can decrypt the information. Figure 10.4 illustrates the process of asymmetric key encryption.

Encryption and Decryption Algorithms

Two algorithms are widely used for the public-key encryption: the Rivest, Shamir, and Adleman (RSA) algorithm and the Diffie and Hellman algorithm. The RSA algorithm randomly generates a very large prime number for a public key. Then the RSA algorithm relates the public key to another very large prime number, which functions as a private

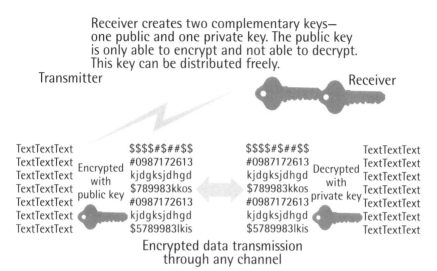

Figure 10.4 Asymmetric key encryption.

key. The integrity of the algorithm is based on the difficulty of finding the prime factors of a large integer. The Diffie and Hellman algorithm is based on the difficulty of computing discrete logarithms in a finite field generated by a large prime number.

Pretty Good Privacy

Pretty Good Privacy (PGP) freeware is a public-key encryption/ decryption program that can be downloaded from www.pgp.com. Commercial versions are available. The program will automatically create public/private keys and key rings and will distribute public keys through public servers on the Internet. It supports all major e-mail packages and allows encrypting and decrypting of data and files. PGP supports the Rivest, Shamir, and Adleman (RSA) algorithm and the Diffie and Hellman algorithm.

Different software vendors offer at least one of these levels of encryption for sending e-mail or documents within an intranet or over an Internet or extranet. The benefit to encryption is preventing would-be snoops from reading data by using a sniffer or other tool that can capture the data as it passes over the communication link or while on the server.

Authentication and Access Control

Authentication is used to identify users and programs and to grant access levels according to information on the network. Authentication also assures accountability. Because activity tracking is a standard tool featured in most corporate intranets and extranets, activity tracking such as file access, modification, and deletions can be set to track according to user login names and passwords. Login names and passwords are one of the most common methods of ensuring authentication, because security breaches can be tracked to a central point of contact—the person logged on to the network.

Authentication can be established by requiring a single login to access information on the network. This is the most common method of gaining access to resources, computers, and data on corporate systems. In more secure environments, double or even triple authentication methods are required for access. For example, in some environments, double authentication, or requiring two individual logins, may be required to access extremely confidential information. With Domino software you can set multiple authentication requirements when adding new servers or users to an environment. Two or more people need to enter, for example, their user names and passwords prior to allowing these additional users into the environment.

Authentication technology may rely solely on logins and passwords. Login names and passwords are used primarily by employees to access network resources. More advanced methods are used to access sensitive data or servers. Smart tokens; smart cards; biometric measuring devices for thumb prints, palm prints, and voice patterns; keystroke recognition; and even retinal scans can also prove effective. Based on the company's security needs, many variations can be exploited.

Digital Signatures for Authentication

The digital signature replaces the handwritten signature as certification of an individual's identity. Signatures provide information such as proof of origin, proof of submission, and proof of delivery. Digital signatures can be used when sending electronic mail or documents on the Internet. These signatures authenticate the author/sender of a document. Authentication proves critical because e-mail messages and electronic documents are more susceptible to forgery or misrepresentation than other means of communication. However, because digital signatures can be based on Social Security numbers and other possibly attainable personal data, they have encountered a certain level of scrutiny from online users and the media. Digital signature technology is still progressing.

Viruses

A computer virus represents one of the most serious threats to the security and integrity of a computer system. Viruses damage systems by deleting files, modifying security, and changing hardware configurations, leaving some systems unusable. Computer viruses spread from user to user, program to program, and system to system. Due to this rapid growth pattern, a virus might be detected too late.

Warning Signs

Some viruses are obvious, while others remain difficult to foresee. The Happy99 virus, for example, displayed fireworks on the screen. The ILOVEYOU virus displayed few symptoms while it destroyed jpeg image files and replicated madly throughout e-mail systems. Basic symptoms of virus infection can serve as a cautionary warning. Although they are not always an indication of infection, these symptoms should raise your suspicion. Potential symptoms include:

- Time stamp or length of files changing unexpectedly.
- Program load time increasing.
- Unexplained writing to write-protected media.
- Amount of available memory on a computer decreasing unexpectedly.
- Executable files modified or deleted.
- Workstation rebooting at random.

How a Virus Works

A computer virus pursues two primary objectives for its creator. First, the virus copies itself into programs or files on a system. Second, the virus executes the instructions that the creator of the virus incorporated. This execution may be triggered by a date, a time, or execution of another program within the computer system.

The virus instructions may include commands to display messages on the user's screen, format disks, oddly reboot the computer, and create other not-so-pleasant effects. Viruses target areas such as executable files, program load files, device drivers such as video, and, in some cases, the Microsoft Normal.dot template.

How do these viruses enter a system? Diskettes traditionally represented the most common vehicle for intrusion. Hackers now introduce viruses via e-mail attachments, software purchased from commercial companies, malicious employees, Internet sites, and other electronic sources. Most of these viruses are aimed at infecting workstations.

Keeping Them Out

Preventing viruses from entering your company's systems can be addressed systematically. The initial step requires policies that dictate security measures. Reduce the number of viruses that intrude. Minimize the damage caused by those that you discover. Incorporate the guidelines that follow as corporate policies.

Back up data regularly so that when a virus does hit the company, all of your data is not lost. In environments where users store most of their information on local computers, ensure appropriate backups. Some companies require that employees place all of their files on a network shared on a central server to help ensure daily backups.

Scan all backups prior to restoring information so viruses are not reintroduced into the system.

Do not allow software other than corporate software to be installed on computers within your company. Often employees carry programs from home and install them on their workstation or download applications and application upgrades from the Internet. Ensure that policies prohibit this activity, or outline which downloads are appropriate and when.

Teach user awareness of potential virus risks and how to minimize possible infections. If people are aware of methods to keep viruses out of corporate boundaries and how to minimize the damages if a virus does appear, they will understand their responsibility in your environment. Users should recognize the signs of viruses. Institute a procedure for reporting potential threats or suspicions of viruses.

Use virus protection software to detect known viruses. It is also important to download or install updates continually. Remind your employees to download updates every few months, if not more often. Several vendors offer virus scanning protection and detection software. Some of these vendors include Data Fellows F-Secure Anti-Virus (www.datafellows.com), Network Associates/McAfee (www.avertlabs. com), and Symantec/Norton AntiVirus (www.symantec.com).

Auditing an Attack

How do you know if you have been attacked? Tools incorporated into any operating system, firewall programs, and other third-party vendor services enable you to monitor information within a corporate infrastructure. In terms of security, determine who is accessing the network, how files are being changed on the system, and where there are excessive "failed login" attempts. In some cases, tools are just a starting point. You may also need to monitor physical forms such as access to the computer system, electronic data storage media access, and access logs. When conducting audits, weigh your resources, including time, staffing, and money, against the damaging effects of an intrusion. Strive for consistency, frequency, and depth.

Some of the areas to audit include:

- Changes to directories and files.
- System logs for information such as login failures, deletion of system files, and system security changes.
- Activities involving sensitive data.
- Changes in user account access control privileges.
- Possibilities of physical intrusions, including computer access or backup media access.

Monitoring tools allow you to determine whether or not an activity is successful. Examples include successful and unsuccessful logins, file and directory access, use of user rights, security policy changes, and system restart and shutdown. If there are consecutive unsuccessful attempts by users, there may be a problem. In addition, several operating systems and applications deny access to a user for a specified amount of time if login attempts have failed beyond the prescribed limit. These systems can also be set to require the administrator to reset the password when failures occur.

CERT

The Computer Emergency Response Team (CERT) represents one of the main agencies for Internet security. This organization was formed by the Defense Advanced Research Projects Agency (DARPA) in November 1988. CERT's mission is to work with the Internet community to facilitate its response to computer security events involving Internet hosts, to take proactive steps to raise the community's awareness of computer security issues, and to conduct research targeted at improving the security of systems.

CERT products and services include twenty-four-hour technical assistance in responding to computer security breaches, product vulnerability assistance, technical documents, and seminars. In addition, the team maintains a mailing list for CERT advisories and provides the Web site www.cert.org and an anonymous FTP server ftp://info.cert.org/pub where security-related documents, CERT advisories, and a variety of tools are available.

Visit the CERT Web site to view the latest security alerts, including viruses and network vulnerabilities. In addition, CERT acts as a central location to report incidents relating to any corporate security problems.

In your quest to secure the computer systems of your company, a comprehensive plan with consistent implementation is your first and best line of defense. Your goal is to prevent a hacker from entering or affecting your network without limiting the communication between your employees and your customers. If you lack inspiration, take another look at www.alt2600.com, which offers descriptions of hacker attacks and their weapons of choice. If you lack direction, review the policy and procedure instructions throughout this chapter.

Use any of the security measures available to you: user procedures, access restriction, passwords, firewalls, proxy servers, encryption, digital signatures, and audits. Determine which measures serve your goals by weighing their benefits against the cost of implementation and maintenance. Keep in mind that, in the end, your security is only as good as your governing policies and procedures. Enforce them so that everyone is aware of risks and remains accountable. Hardware and software deployments alone are insufficient. Make the commitment to secure your company.

11

Get Ready for Wireless!

Few technologies in the history of technology have generated the promise, potential, and excitement of wireless communications. The twenty-first century bridge between our perception of science fiction and the reality of science and technology is being created and developed at breathtaking speeds. Mobile communication is beginning to affect the daily lives of a huge number of people worldwide. More so than the advent of the personal computer and the World Wide Web, wireless communications technologies and applications will penetrate every aspect of our lives. This not only involves data retrieval, electronic commerce, messaging, and Web access; it also will create the ability to send and receive wireless signals to control machines and devices.

To maximize your benefits and prepare yourself for the wireless future, it is important to know more about the technology being developed and the pace at which it is approaching. A working understanding is critical for your future success, whether you are a consumer, a businessperson, or an information technologist.

To truly understand how wireless technology will affect you, it is necessary to see the big picture. Although this book is centered on effective e-commerce solutions for your business, the advances in wireless technologies will influence far more than new avenues for marketing and selling products. In essence, the wireless technologies we talk about in this chapter will change the way in which people carry on the daily aspects of their lives. The design and use of products will influence how and what people buy. So if you want to truly understand your potential for the future, you must put yourself outside the box and consider the entire future world of wireless communications.

In this chapter we will cover some of the basic aspects of wireless communications in the context of how they might affect your life and business. Because of the explosive changes happening in technology and applications, before this paragraph is finished the world will have advanced beyond what we talk about. The key to success in the wireless world will be to keep abreast of the changes and advances and to realize that the ground rules of behavior for this new world are new to everyone—and you have the unique opportunity to help establish those rules. The next five years promise to be a time of extraordinary excitement. Enjoy the ride.

From E-Commerce to M-Commerce

With every new technology comes the inevitable buzzword or catchphrase that quickly becomes part of our everyday conversation and language. Because the primary goal of most companies is to capitalize on wireless technologies and services to make money through transaction-based applications, many involved in this aspect of business are calling this technology mobile electronic commerce or *m-commerce.*

Only those just arrived from another planet will not have noticed the worldwide proliferation of mobile phones and handheld wireless communication devices. The world has leapfrogged the required access to the Internet and voice communication through wired telephony. Mass quantities of wireless communication devices can now be found in every corner of the globe. Social scientists are labeling these widespread communication devices as the world's best hope for future world peace, because the instant flow of information creates a level playing field for all.

With the ability for a mass of people to achieve instant communication also comes the increased consumer need for instant gratification from these same devices. People want their information, and they want it now. Satisfying this need, whether for information, for data, or for m-commerce, unleashes unlimited possibilities for the entrepreneurial businessperson.

How Many Customers? How Soon Will This Happen?

In researching the statistics on e-commerce users for the first and second editions of this book, it became clear that consulting firms are very conservative in their estimates. Each month, five- and ten-year projections have been released with greatly increased projections over the month before. This may also be the case with projections for m-commerce users. Take a look at Figures 11-1 and 11-2, which display the number of current and projected wireless communication device users in the United States and worldwide over the next six years and the number of the total users who are expected to engage in some form of m-commerce, respectively.

The chart in Figure 11-1 is the estimated number of people subscribing to wireless portals in the United States, as given by The Strategis Group. If the projections shown in Figure 11-2 hold true, this number may be greatly understated.

The figures show that most analysts expect that approximately 25 percent of wireless users will engage in some type of m-commerce on their mobile sets. Given the vast array of commercial applications that are being developed in all corners of the world, this may be a great underestimate. Suffice it to say that with an anticipated base of two billion wireless customers by 2005, with 25 percent or more engaged in m-commerce, the numbers are very impressive and very inviting for the creative entrepreneur.

The financial firm Merrill Lynch estimates that $25 billion will be spent worldwide by 2005 on wireless software and m-commerce. If the projections of two billion are correct for the number of wireless device users worldwide, and if 500 million users are engaged in m-commerce, then it appears that the estimate of $25 billion could be understated if only $5 per user is anticipated to be spent. Needless to say, it's a large pie, and we should expect a large number of businesses and people to want a piece of it.

The Devices . . . Now, Then, and in the Future

When we talk about wireless devices, our speculation on the future of these machines is only that, speculation. We assume that no matter what happens in the development of future hardware, software, and

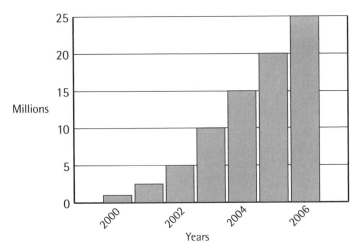

Figure 11.1 U.S. wireless portal subscribers.

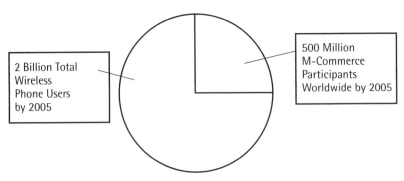

Figure 11.2 Of the 2 billion worldwide wireless phone users by 2005, 500 million users will engage in some form of m-commerce.

applications, one truism will remain, and that is that the device must be portable. Whether you are a fan of wireless phones, PDAs (personal digital assistants), pagers, or palmtop computers, you can expect the designers of the world to concentrate on portable devices that can easily fit in the purse, in the pocket, and, most importantly, in the hand. Although wireless technologies will be used in a myriad of devices, such as General Motor's On-Star system, or in remote-controlled appliances, the overwhelming priority will be the development of personal portable communication tools.

The challenges associated with developing handheld portability are enormous and are not to be taken lightly. The number of people and the number of companies involved in the design of hardware, software, middleware, and communication systems to overcome these challenges are considerable. Compared to large traditional engineering projects, it would be safe to say that the total R&D expenditures, the total number of people, and the total number of work hours being invested in wireless technologies far exceed the total input in the worldwide space program of the last forty years.

In the "old" world of e-commerce, on the Internet through phone lines, the physical design of your terminal was not critical to how the machine was used. Of course the speed of your modem, the quality of your monitor, and the size of your computer memory contributed to the quality of your e-commerce experience. However, you probably sat at the computer and typed and used your mouse to conduct your shopping. Consider the difference in potential m-commerce scenarios when you are carrying your device. This creates a new business paradigm where the human–device interaction, coupled with the design of the device, helps determine the vast array of potential uses for m-commerce. We will cover this in more depth in the following sections, but consider the possibilties created by the mere fact of having a portable device.

Mobile Phones and PDAs

According to a study by The Yankee Group, by the year 2003 more people will utilize wireless communication devices to go on the World Wide Web than will use computers. They go on to report that the world already has more people using wireless phones than the total number of people using the internet. And in four years, twice as many people will be using wireless phones than will use the Internet.

As mentioned before, the surge in the sale and use of digital wireless phones over the past two years has been spectacular. We are bombarded with advertisements selling phones and services. Or, considering a very unusual business model, selling services and giving away phones or selling phones and giving away services. No matter how they are bought and sold, we know they have quickly been absorbed by and merged into our culture.

In addition to wireless phones, PDAs or personal digital assistants and palmtop computers have gained widespread use in business and other settings. Developed simultaneously with, yet independently from, mobile phones, the goal and long-term business plans for each provided for a future involved in m-commerce.

What's Available Now in E-Commerce?

The list of currently available wireless services available on mobile phones and PDAs is limited but expanding on a daily basis. Although the services may not be as robust, fast, or efficient as similar Internet services, all groups are attempting to improve the quality and quantity of information delivered.

You can now get reliable up-to-date sports scores, news, and weather reports, as well as access to a limited Web environment. Services that list restaurants and shops, as well as discounts and other e-commerce promotions, are expanding. PDAs and similar devices have also advanced the ability to edit and send data, word-processed documents, and spreadsheets to other sites. These devices also allow you to interact directly with your personal computer to exchange and store information. In addition, limited direct m-commerce is starting to be a real option. Advancing very quickly is the ability to track, buy, and sell stocks and bonds via a service through mobile phones.

Current physical device limitations, coupled with low communication transfer speed, add to the challenges facing wireless technologies and m-commerce. However, where there are large profits to be made, solutions usually follow. In later sections in this chapter we explore some of the possibilities. Unlike science fiction, most of what we discuss is not only possible but probable and a lot closer than you may imagine.

What Will Be on the Screen?

Consider your personal computer and the amount and quality of information and visual images you can see on your screen. Now, for those of us old enough to remember, think about the look and feel of the pioneering computer game "Pong" that was released just over twenty years ago. Who could have imagined that the progress in technology in such a short period of time would take us from "Pong" to the World Wide Web, to DVDs, to MP3?

It is not out of the realm of possibilities to use the prior paragraph as an analogy to the difference between the current state-of-the-art in wireless communication devices and those being developed for the future. Except for the ability to carry around a seventeen-inch full-color display screen, you may expect your wireless device to meet the abilities of your current PC. (Of course, with advances in projection displays and holographic imaging, someday a big display may be possible.)

Right now, the display quality and the bandwidth do not allow an easy transition between how you currently create, store, retrieve, and view Web-based material on a wireless device. As users, we may have to change our perceptions and desires when using these devices; however, the device designers and engineers are doing everything possible to please the consumer. We can't guess at the future, but we can anticipate the progress. Later in this chapter we will examine some of the current technologies of other devices in anticipation of the possibilities of the handheld wireless devices of the future.

Mobile devices are playing the central role in communication is today's world. An extension to the Internet is just logical. New generations of wireless devices will provide us with data and information residing on the Internet, not only in the form of pure data but also in the form of music, video, and interactive media communication. The key component of wireless communication is WAP (wireless application protocol).

News services such as CNN Mobile provide news and information service directly to cell phones and wireless devices. Generic wireless portals will allow interacting with different web services. MSN Mobile (http://mobile.msn.com/), for example, allows the user to read e-mail through Hotmail.com and to browse MSNBC.com, to receive news, sports, and weather updates. MSN allows allows the user to review travel and flight itineraries through Expedia.com and use MSN Yellow Pages. Yahoo Mobile (http://mobile.yahoo.com/wireless/home) allows access to a wide range of services such as Yahoo! Auctions; mail forwarding and messaging; and WAB directory services, which feature vendor solutions and mobile alerts for sports, stocks, weather, and horoscopes sent to cell phones or wireless devices.

Wireless Portals

Perhaps the most important aspects in the future of wireless communication and particularly regarding m-commerce is the creation, design, and control of *wireless portals*. A wireless portal, as defined by The Strategis Group, is a customized point of entry through which a wireless subscriber can access Internet sites and information.

The best examples of wired portals are currently AOL and Yahoo! They have established a strong market model based on a one-door entry. It appears that consumers prefer the ease of going through one site (a portal) to reach the majority of their information, e-commerce, and communication needs. Even if you are going to a consumer Web site to purchase something directly, it is likely that you will log on through an internet service provider such as AOL.

The assumption by most industry experts is that consumers will come to expect a similar single-entry wireless portal to meet their needs. Early surfers of the Web will remember the difficulty of using one company as an ISP (internet service provider) while still needing to purchase other software pieces, such a search engines, in order to reach some level of functionality on the Web.

Another aspect of single wireless portal access that should not be underestimated is the consumer preference for a single payment and billing site. This appears to be the desired method because of convenience and control, although not necessarily of savings. When AOL switched the majority of its customers to single-rate monthly billing, many were skeptical of its possibility for success. The question was, how many people would be willing to pay the fee when the hourly rate of the vast majority of Internet users was lower than the monthly fee? Evidently, most people are willing because of the strong perception of controllable billing.

This is a very important factor in considering m-commerce. Will customers prefer to have their unified wireless portal service provider also be the one source for billing? If you become a frequent buyer from different stores, would you prefer the convenience of the billing and payment to be done through your wireless portal provider? This may seem a bit irrelevant at this stage; however, it is one of the most important elements in the fight to become the winners in the portal wars.

Consider the fact that you can expect the cost of wireless portals to be charged back to you in a variety of ways: by subscription to the service, through service charges from your wireless communications carrier; and/or by commissions paid to your wireless service provider. With an estimated 500 million people taking part in m-commerce by 2005, you can only imagine the huge amount of money at stake.

No one is certain who will dominate the wireless world's version of the ISPs, but let's take a look at some of the current elements.

Who Are the Players? Carriers vs. Internet Service Providers

With such huge potential for revenues and customer access over wireless communication devices, the competition between companies for a share of this market is tremendous. In a recent article in *The Washington Post,* Leslie Walker summed it up well by saying, "It is the second great land grab of cyberspace, the one for control of the wireless Web. Phone carriers are competing against and partnering with Web portals, device makers, and a new class of mobile service providers in pursuit of revenue from mobil e-commerce."

> *"The second great land grab of cyberspace . . ."*

During the last five years we have all been witnesses to some of the most interesting business phenomena in history, from the huge amount of wealth created in the computer and Internet world to the vast amounts of cash gained and lost on "dot-com" investments.

Few scenarios are as interesting as the rise to dominance of companies such as AOL in an arena that we might have expected to have been dominated by the communication carrier companies. After all, the carriers owned the telephone lines, and they had direct access to the customer base. All the carriers needed to do was design and deliver a portal for their customers. But, by the time the carriers realized that this business even existed, it was too late.

Both the management of the telecom carriers and the stockholders of communications stock are determined not to let this happen in the wireless world. With wireless communications and particularly wireless portals up for grab, the carriers think they have another unique advantage. In this case, they not only control the lines or wireless airways through their subscription services, they also are closely tied to the control and development of the wireless devices.

The carriers are also counting on recapturing m-commerce users because of consumer confidence in the phone company as a billing and payment organization. Would a customer feel more comfortable paying for an m-commerce purchase through his or her phone bill rather than through an ISP or directly to a business? Perhaps yes, given that a utility company may give a customer a sense of security of information as well as the added convenience of single billing.

To push their version of m-commerce ahead, the carriers are engaged in a huge array of partnerships with hardware and software companies, as well as with information providers and businesses in order to build wireless portals.

However, do not underestimate the power of the ISPs in capturing the wireless portal business. After all, what is the one service that will be offered on wireless portals that has become the most important aspect of the internet? E-mail! It is something most people can no longer live without. For those of you who have tried, how difficult is it to change your Internet e-mail address? With a pool of hundreds of e-mail contacts, this is no easy task. Anne Bentley of AOL, while talking about the ISP fight for the wireless portal market, said, "We're banking on the fact that we've got 23 million customers who are used to AOL services and will not want a separate e-mail address" on their mobile phones.

In addition, the AOLs and Yahoo!s of the world are structured to design and adapt user interfaces that meet the needs of their customers. With such a large market share, consumers may expect the look and feel of Web-based sites currently available form ISPs. Don't expect the carriers to capture or take away the brand loyalty of ISPs.

Also, as one report expertly pointed out, there is nothing that could prevent AOL from selling and leasing mobile phones and devices. In the wired world of the Internet, this would have been impossible. However, given the unrestricted boundaries of the future of wireless, coupled the size and market clout of AOL, this is not out of the realm of possibilities.

What Will Happen?

It is important as a businessperson and consumer to understand the battle for the wireless portal market being waged between carriers and ISPs—not because it will directly affect your business in the future, but because this competition is pushing the introduction of unlimited m-commerce services far faster than they normally would develop. If you wish to be a player in this current and future m-commerce market, you will benefit from following these developments.

Overall, most experts do not expect either the carriers, the ISPs, or any third party to dominate and control your Web portal. Most likely it will be a combination of a variety of companies who engage in what they do best. The design and use of Web portals may lean toward the expertise of ISPs and the control of devices and the airways toward carriers. It is anyone's guess at this point, but it will no doubt be exciting.

At this point, let's look at the developments in the international market outside the United States. Certain factors in the development of the Internet, coupled with diverse social and industrial structures, may make Europe and Asia the models for the future of the wireless world.

International Perspective

If Americans believe that the use of mobile telephones is exploding across their country, they must understand that introduction of wireless communications in the rest of the world is more like a nuclear explosion.

From the sophisticated cities of Europe to the bustling markets of Asia, mobile phones are everywhere. As we showed above, the market penetration of mobile phones may reach two billion sets worldwide by 2005. With the vast majority of users outside the United States, we expect that the major influence on both wireless portals and m-commerce may come from Europe and Asia.

We have discussed the competition between carriers and ISPs for the burgeoning Web portal market. Although this is happening now, it is primarily a fight being waged in the U.S. markets. With the great population of users outside the United States, developmental factors, particularly in Europe, may have a greater influence on worldwide m-commerce than the carrier/ISPs of the United States.

Although AOL Europe has strongly developed a European base, it in no way has been as influential on the internet users of Europe as in the United States. The same can be said for electronic commerce on the Internet. More people in Europe and Aisa use mobile phones than the Internet. Most likely, customers' first exposure to e-mail and e-commerce will be on a wireless device. Therefore they will not have the built-in brand loyalties or the expectations of the human–computer interaction of ISP-type portals.

Without ISPs playing a central role in the daily computing life of Europeans, the cooperation between carriers, device manufacturers, information providers, and merchants has been greater and more widespread. Manufacturers such as Ericson and Nokia in Scandinavia work directly with consumer-oriented companies to develop products that deliver what the customer wants.

The development and the acceptance of wireless products depends on the underlying culture and lifestyle of communities and countries. The future of wireless portals and subsequently of m-commerce depends on pleasing the customer. Few people would disagree that Europe has dominated the world of industrial design. From Sweden to Italy, the design of furniture, appliances, automobiles, and clothing has created a culture of responding to the needs and desires of the consumer. Expect this to continue with the design of Web portals.

In addition, in Europe and Asia the populations are far more densely located and subsequently more urban. This creates two reasons why these markets may become more influential on the development of Web portals and wireless devices. First, the density of the market makes it easier to advertise and evaluate new products. Secondly, changes and new products can be absorbed by a larger percentage of the population than in less urban markets. This is an enviable consumer group for any supplier of information, goods, and services.

One other factor that should make the transition to m-commerce easier in Europe is a far greater acceptance of electronic transfer of funds. In many countries in Europe the telephone companies were created as public utilities, often under the same management as the postal system. In addition, the banking system was closely tied with each service. Long before debit cards were introduced in the United States, Europeans were paying utility and other bills electronically through their banks. In addition, debiting an account to pay for restaurant bills and goods is quite common.

On the other hand, in Asia, where the majority of the population could never afford wired telephone service, mobile phones are everywhere. This leaves open a number of paradigms and business models where customers have no experience with either a carrier or an ISP. The reaction of these consumers on the look, feel, and usability of Web portals will be pure and objective and not based on any preconceived notions. What an excellent market for testing products this will make!

Keep an eye on developments internationally in order to follow what is best for your business. In the early part of this chapter we said that wireless communication has the opportunity to contribute greatly to world peace because of better interpersonal communications. The same can be said for world trade. Don't assume that because the United States has dominated the wired world of the Internet that it will control the future of the wireless world. The more open and knowledgeable you are of worldwide advances, the better your chances of success in the world of m-commerce.

Customer Satisfaction

A handheld wireless device is the ultimate tool for the "me generation." Along with a society that needs instant gratification comes the need for instant satisfaction. With the mobile/wireless environment, the customer is already "walking," and he or she won't hesitate to walk away if things don't function.

> *The customer has never been more right than he or she will be in the world of m-commerce. . . .*

Therefore the customer has never been more right than he or she will be in the world of m-commerce. This creates quite a challenge for everyone involved in the wireless world. The carriers, the wireless portals, and the device manufacturers, as well as the m-commerce merchant, must realize that the very technology that has brought the customer to them can also lead the customer to another supplier.

The idea of "throughput levels," or the success of downloading information, can be extended to the customer. Customers who do not see or receive what they expect will go elsewhere. Because much of the display and communication technology is still in rather rudimentary stages, one should also expect that the charges for services and devices should be low until the quality is there.

To the merchant, this may mean little or no profit in the short run. The goal should reflect the need to gain and keep customers, particularly those customers who will help you learn from their experience.

Early ISPs suffered very high rates of "churning" (signing up and subsequently dropping a service) compared to current mobile carrier churning rates. In order to survive, each business must have a plan to monitor and respond to customer feedback to prevent churning.

Wireless Internet Convergence: Merging Technologies

A vast array of technologies and applications have been developing for nearly twenty years. Not only have people and businesses become accustomed to using devices and data in new efficient ways, the manufacturers of devices, the middleware/software developers, and the communication companies have been increasing the offerings and capabilities of available goods and services. As you can see from the list below, many of these technologies developed from totally independent sources with independent needs and goals. However, the capabilities and possibilities are now intersecting at a record pace. Throughout the next five years a huge number of applications that people have not even considered will arise. You have the opportunity both to use these applications and also to contribute and gain from creating your own possibilities.

The following list shows a brief scan of some areas that are taking advantage of wireless applications. It is not important to dwell on the specific devices for the purposes of this chapter. They are presented here to give the reader an idea of what is currently happening. You can safely assume that all of these applications will be developed to work on the smallest of portable devices or at least interface with small wireless devices.

1. *Portable work devices.* These are already being seen as UPS drivers' tracking systems and rental car company wireless handheld terminals for car returns.

2. *Transportation management.* Trucking companies utilize laptop computer systems to monitor, track, and manage every aspect of their drivers' and vehicles' daily activities, including the driver's log (sleep, hours of driving), fuel prices, weather and road conditions, parking availability, load tracking, speed, and gas efficiency, as well as complete company-driver communications.

3. *GPS and mapping devices.* Global positioning devices let you know, through satellite tracking, where a device is located. In addition, wireless availability of maps and charts allows the user to plot directions.

4. *Personal vehicle systems.* Systems such as the highly promoted On-Star system from General Motors combine wireless communications and GPS tracking with live communication to a customer service person to give a vast array of services.

5. *Inventory and ordering systems.* For years the grocery and the parts businesses have used direct wireless systems to take inventory and automatically order. In addition, new technologies in cash register sales/returns tracking provide instant wireless communication between retailers and suppliers.

6. *Banking and finance.* Banks and other financial institutions are leading the way in offering services available through wireless devices as well as sending fund transfers through wireless networks.

7. *Utilities.* For many years the power companies, phone companies, and cable providers have used wireless devices both to analyze on-site problems and also to communicate with centralized information sources.

8. *Toll roads.* Driving through toll booths using the "e-lanes" allows you to save time by paying tolls through a wireless communication account.

9. *Science.* Field work and laboratory work can now combine notes, GPS data, and computational power as well as instrumental data, all through wireless networks.

10. *Palmtop/PDA applications.* These available devices combine a vast array of functionality, including calendars, clocks, address books, e-mail connectors, Web searches, database communications, limited searches, and developing products such as electronic books.

11. *Military.* In the military, wireless devices are being developed for everything from team communication, to field medical guidance, to target tracking, to information sharing, to troop strength management.

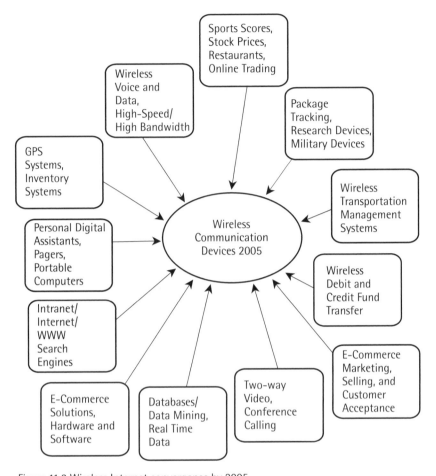

Figure 11.3 Wireless Internet convergence by 2005.

Advancing M-Commerce Applications

We have looked at a number of diverse applications from a wide variety of sources using wireless technologies to enhance their businesses. Let's now examine some advances in wireless applications that you could consider part of m-commerce. This is a short list, but perhaps it will open your imagination.

1. *Point-of-purchase checkout.* Mobile phone manufacturer Ericsson and English food giant Safeway are developing point-of-purchase checkout for customers. Customers will be able to scan the product they wish to purchase at the shelf, place it in their cart, and be automatically checked out with a debit transaction sent to the customer's bank.

2. *Coupon downloading.* Many services now offer discount coupons to products with which you can be at the point of purchase, determine what you wish to purchase, and then download a discount coupon and send it via wireless to the cash register of the store you are visiting.

3. *Marketing and promotion searches.* Do you know what you are looking for, but want to know where to find it? A wireless Web search can let you know where to find it as well as let you know of any sales and discounts available.

4. *Wireless debiting/electronic wallets.* Nokia and other device manufacturers are working with companies to send wireless signals to debit the consumer's electronic wallet or bank account. Imagine using a vending machine without change. Your wireless device sends a message to the machine that gives you what you wish while at the same time debiting your account.

Let's take this one step further. Imagine that in addition to everything else listed in this chapter, your new wireless device can also do the following; open your garage door; start your coffee maker; keep track of your children; control your smart-house devices, such as thermostats and window shades; carry out simple tasks such as opening your house and car doors and remote-controlling your television; and, who knows, even program your VCR. The possibilities are endless.

We could go on and on with ideas, but let's use one case study that is often referred to in the context of many of these applications rolled into one m-commerce scenario. After considering this situation, why don't you see what you can come up with for your future? Don't limit yourself to what is currently happening, but put all of your ideas together to see what you can do.

Case Study: Year 2005

You are walking down the street of a commercial district when the personal wireless device you are carrying signals you that the store you are passing has a sale on brown widgets. The global satellite system senses that you are in the immediate vicinity of the store. You have programmed your personal preferences into the computer, and the device "knows" that you would like to find a brown widget if convenient. At the same time, the device logs into the Internet and does a search for availability and prices of brown widgets and shows that this brown widget is a good deal. Also, the wireless device can download any available discount offers from the merchant or the manufacturer directly into your machine. You proceed into the store, pick up the item in question, and walk out of the store knowing that your wireless device has paid for the item, minus any discounts, and debited your account.

Make Your Plans

If you can foresee your participation in business involving any aspect of the wireless world, from m-commerce, to information provider, to device maker, make sure to start building it as an integral part of your business plan. If you redefine your vision and goals, you can start developing your strategy to deploy the best possible solution. So, unfold your vision and establish a strategy that will meet your business goals. You are limited only by the limits of your imagination.

You are limited only by the limits of your imagination.

Note to Readers

The statistics and information in this chapter are compiled and averaged from data available in numerous published articles, and much of the data is commonly attributed to the outstanding research firm The Strategis Group (www.strategisgroup.com), which is known for its superb tracking and analysis of the wireless world and also from

Forrester Research (www.forrester.com), which we have frequently mentioned as the business world's key resource for information on electronic commerce. If m-commerce and wireless technologies are in your future, purchasing the research reports from these companies may be well worth the investment for your business. Visit the Web sites for these companies, as well as those for Jupiter Communications, Merrill Lynch, The Gartner Group, IDC, The Yankee Group, and others, as well as wireless magazines and newsgroups, to keep abreast of these rapidly changing markets. In order to get a broad picture of information on wireless technologies and markets, try doing an Internet search on "wireless portals." You may wish to try this search online rather than through your wireless device, unless you are prepared for some unusual displays of information.

12

Commit to Never-Ending Improvement

As we look back at e-commerce as it was two years ago, we have seen major shifts in the way companies are doing business online. What was hip then is not what customers want today. Those companies that are successful and staying successful are constantly reshaping and reevaluating their e-businesses. Companies such as Amazon, eBay, and Dell (and the list goes on) have made sure to stay ahead of the curve in the e-commerce arena. Commitment to never-ending improvement is the key to staying ahead.

In this chapter we will look at what a few companies have dome to stay ahead and remain successful in this market. You will notice that it is a combination of offering more and better services, redefining their position in the marketplace, researching customer buying trends, partnering with companies, and continuing to take advantage of technological changes.

The Dell Story

Dell was one of the first companies to reap the benefits of e-commerce. Dell has continually improved its site to include more personalization and one-to-one relationships with it customers. It has consistently increased online customer service and has expanded its business to provide added value services. Currently, more than 40 percent of the company's revenue is from sales online. This year it has generated approximate $15 million per day of revenue online. Dell expects to generate half its revenue online by the end of 2000.

Dell attributes its success to customer service and providing a personalized experience for its shoppers. In 1999, Dell was named one of the winners of the Ten Best Web Support Sites from the Association of Support Professionals.

Although Dell is one of the leading e-commerce success stories, it has not stopped thinking ahead. This year it invested $26 million into SiteSmith, an application service provider that is expanding its services worldwide. You may ask, why would Dell go into the application service provider market? Well, Dell's acquisition is part of the company's efforts to expand into new markets and increase its revenue stream beyond just selling computers online. Dell's goal is to increase its infrastructure services using the ASP business model.

Another initiative is the development of Dell Ventures, which will focus on making strategic investments in early-stage private companies. Dell is also offering clients value-added services such as Web design and e-commerce storefronts.

The eBay Story

eBay, with a simple concept (Web-based auctions) and a market capitalization of $16 billion, has harnessed the resources of the Internet to capture over two million registered users. Has it stopped looking ahead? No! eBay is constant looking for ways to increase its markets by providing international sites and moving into new markets.

This year, eBay partnered with zipReality.com to provide a new category of products and services called eBay Real Estate. zipRealty.com will provide professional real estate services as part of the new category. By taking advantage of zipRealty.com and the powerful technology and infrastructure of eBay, both companies will win in this market. Even though eBay is a brand name for auctioning online, it still keeps moving forward and thinking of the next steps it must take to stay ahead of the market.

The Amazon Story

Amazon is another branded company on the internet known for selling books. Although Amazon is selling other products, such as lawn, patio, and kitchen products, it is still known for providing quick and reliable delivery of books. Amazon is one of the online companies that is constantly improving the way it does business. One of Amazon's key innovations is the 1-Click ordering.

Once a customer has made a purchase online or registered, they can select the 1-Click ordering. This will automatically select all of the previously entered billing and shipping information. Amazon realized that its market was based on convenience and impulse buying. So, it was extremely innovative in using technology, which allowed customers to bypass all of the billing and shipping forms and focus more on making the one click.

The JCPenney Story

JCPenney.com has generated over $79 million in sales online through June 2000. The authors initially met with JCPenney over two years ago to talk about e-commerce and how it is changing the way companies are doing business. It is amazing how JCPenney went from including a few catalog items online to becoming one of the most visited sites on the Internet.

In order to move ahead, it formed a separate subsidiary called JCP Internet Commerce Solutions, which focuses primarily on its e-commerce presence and catalog. The key to JCPenney's success was thinking about how it could provide the best customer service, fulfillment, and personalization online.

We are seeing that often brick-and-mortar companies will create a new department or company purely related to e-commerce. Why? Because e-commerce is not completely the same as selling via catalog or in a brick-and-mortar store. Customer service, personalization, sales and marketing and other areas are addressed differently. By creating a new division or spinoff company, the company can bring in the right resources and experience. Yes, you can retrain people within the company. However, it takes time to shift a large corporation to think in e-commerce.

The Bid.com Story

Bid.com was a company that created an auction-based site focused on business-to-consumer auctions. It wanted to provide a site similar to eBay. The results—it lost substantially over the year. In the spring of 1999, the company's share price crested at $30.70 in Toronto but then suffered a long decline, bottoming at the end of May at $2.80 per share.

Did Bid.com keep on pushing its site to the business-to-consumer market? No! Instead it rethought its business model and shifted toward the business-to-business model. The company knew it had a great auction technology, so it moved to a business model based entirely on selling the online auction technology and services to other businesses. The company business revenue model is now based on implementation fees, monthly hosting fees, and transaction fees from the businesses.

Bid.com is now seeing some successes with this new model. Customers such as GE Capital and strategic partners such as PricewaterhouseCoopers are becoming part of Bid.com. In this example, we see a complete shift in the offerings from an e-company. Keep in mind that the e-commerce market is fairly new. Companies will need to test what works and what does not work. The key is to do it right once. You may find over time that what you thought would work does not work. For example, you may think that selling XYZ products is your company direction. Over time, you may realize you are strongest in distribution and customer service and very weak in the product sales area. So, you think about providing distribution and customer service for other e-companies instead of focusing on selling products.

Think! Think! Think! What's Next

We have covered the way a few companies are thinking ahead. Because the e-commerce field is constantly changing with innovative individuals and companies moving ahead, never-ending improvement is the key.

The following is a great list to help you make sure you are not forgetting the areas to focus:

I. Read.
   a. Know what the latest research/inventions/tech solutions are.
   b. Know what is happening in the market around you.

II. Research the trends.
   a. Your customers change daily—do you know how they've changed?
   b. Available products change daily.
      i. Do you know what's out there?
      ii. Can you add new products to your site?
   c.. Your competitors change daily.
      i. Do you know what they're planning?
      ii. How can you position yourself away from them?
      iii. How can you keep your selling point unique?

III. Invest in change that improves access to your goods and services.
   a. Invest in marketing repositioning.
   b. Invest in Web site enhancements.

IV. Invest in change that adds value to your goods and services.
   a. Invest in service upgrades.
   b. Invest in product upgrades.
   c. Invest in information upgrades.

V. Create barriers to entry for subsequent competitors.
   a. Make sure the value of your product/service increases with higher use or distribution.
   b. Keep your branding unique and strong.
   c. Define yourself apart by honing your unique selling point.

VI. Network!
   a. Become involved in groups that feed on success.
   b. Join the e-commerce consortium.

Read

Reading both trade magazines and e-commerce magazines will keep you informed about what is happening in your market from both the business and technology perspectives. When consulting to businesses and in our courses, we stress spending at least 20 minutes a day keeping up on e-commerce technology and business specific in your industry. This will help ensure that you are at least current on the e-commerce market. In addition, it is a great method to come up with

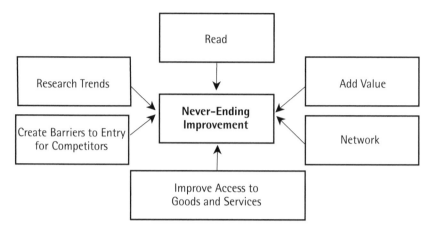

Figure 12.1 E-commerce is an ongoing adventure.

ideas on how to improve your e-commerce solution. Some of the best reading sites for e-commerce are included in Table 12.1.

Research the Trends

You should research both industry trends and e-commerce trends. You should also spend a considerable amount of time researching your own customer buying habits. Most commerce applications come with analysis tools that will help you determine what people are buying, when they are buying, what advertising campaigns were successful, and so on. You may find that certain products are not doing well on your site versus other products. This could stem from a few reasons. Some of them may include the following:

- *Type of product being sold.* Is it an impulse item, such as a book, or is it an item, such as clothing, that customers may be a little wary of buying online?
- *Positioning of item on the site.* Is the item in a location that is difficult to access? Is the consumer being led to this area, or does it require extensive searching?
- *Cost to the consumer.* Are the costs too high—either product or shipping?

Table 12.1   Reading Sites for E-Commerce

www.internet.com	Provides the latest new on internet and e-commerce business and technology trends
www.cyberatlas.com	Provides statistics on demographic and geographic buying trends
www.adresources.com	Provides information on marketing your site, statistics of advertising mechanisms, and new advertising technologies
www.zdnet.com	Provides general e-commerce information for small and large businesses; also has a great section called 10 best and 10 worst e-commerce sites on the Internet
www.ecommercetimes.com	Great site for getting overall trends on e-commerce; additionally, has a section called Success Stories of e-commerce companies
www.entrepreneur.com	Online magazine that contains great information for small businesses in the e-commerce arena; also provides an "Ask the Expert" section so you can get specific information
www.thestandard.com	Site filled with great information on company profiles, e-commerce mergers, and other information on e-commerce

Also, make sure you know what your competitors are providing on their sites to attract customers. They may be providing incentives such as coupons or special discounts for the same items. They may also be providing additional services that you may not have incorporated. You do not want to look exactly like your competitor. You must still differentiate yourself. However, you should know at a minimum what they are providing so you position yourself away from them and keep your selling point unique.

Invest in Change That Improves Access to Your Goods and Services

Think about how you can improve access to the goods and services on your site. Are you incorporating all possibilities for marketing to include both traditional media and online media? Have you checked which are working and not working to your advantage? This will allow you to determine where to place your marketing time and dollars. Why spend thousands in one area if it only brings you a minimal amount of business?

Once the customers get to your site, are you making impulse buying easier? Are you providing information on your site to help customers hit the final "Pay Now" button? Does your site reflect your current and potential customers as far as site design, look, and feel?

Invest in Change That Adds Value to Your Goods and Services

Are you constantly thinking about how you can add value to your customers' buying experiences? Are there other servers, products, or information that can be provided. Think about Dell, which keeps on thinking about the next steps and how to add value.

Create Barriers to Entry for Subsequent Competitors

Always stay one step ahead of your competitors or, better yet, two. Constantly think about what's next, how you can improve customer service and sales and marketing. What do you have that your competitor does not that would increase market share? It may be something like strong relationships with your suppliers who in turn may allow you to make special deals such as direct distribution to the consumer or marketing revenue. It may also include a unique way of selling a product or providing customer service, like the 1-click offered by Amazon.

Network!

Go to trade shows; show your products and services. Networking is one of the best methods to keep updated. You may want to start attending trade shows of potential customers to find out what they would like to see on your site. At the same time, you are building relationships. Think about all of the marketing and relationships you will create.

In addition, by attending e-commerce trade shows, you will be able to directly communicate with commerce vendors and consulting groups that can help put your site together or improve your current functionality.

This is a time of remarkable opportunity for those businesses that harness the power of the new market . . . e-commerce.
Enjoy the ride!

Index

ISBN 0-12-421161-5